# Judging Extreme Weather

Written by the World Meteorological Organization (WMO) Rapporteur of Weather and Climate Extremes, this book addresses the reality of extreme weather—how it occurs, how we measure it, and what it means for our future.

Weather affects everybody, and with the increasing impact of climate change and the prevalence of storms, droughts and floods, it is clear that we are affecting all aspects of weather. Consequently, people love to talk about weather, complain about it, argue about it—and be intrigued by it. Twenty-four/seven coverage of the weather, however, has helped foster a tendency for marked overstatement—the creation of misconceptions, exaggerations and, frankly, even outright lies. Leading expert in weather and climate, Randy Cerveny, draws on his extensive experience with the WMO and personal research to give the reader a behind-the-scenes look at how weather and climate extremes are recorded and defined. He unpacks the science behind these extremes through a number of specific WMO investigations that span a diverse range of countries and weather events, including lightning, rain, hurricanes and tornadoes. Cerveny balances these factual accounts with playful interludes that detail bizarre and intriguing weather-related stories and anecdotes.

This compelling book is a must read for all those interested in the science behind extreme weather.

**Randy Cerveny** is a President's Professor in Geographical Sciences who specializes in weather and climate at Arizona State University. He obtained his doctorate from the University of Nebraska in 1987, has studied weather around the world, and has stood on all seven continents. His research has ranged from studying the weather associated with prison escapes to weather of the far future. His research has garnered the attention of the BBC, CNN, ABC News, NPR and others. He was an expert for the Weather Channel and currently appears on the National Geographic's show *What on Earth?* His first book, *Freaks of the Storm*, was published in 2006 while his second book, *Weather's Greatest Mysteries Solved!*, was published in 2009. Since 2007, he has served as the World Meteorological Organization's Rapporteur of Weather and Climate Extremes, the person in charge of assessing and validating world extremes, such as world's hottest temperature.

"Professor Cerveny is known as the authoritative source for confirming the extreme weather – climate records. In this book, he provides compelling behind-the-process insights and jaw-dropping facts about some of the world's most extreme weather."

**Marshall Shepherd** *is the past president of the American Meteorological Society, Fellow of the US National Academy of Sciences and National Academy of Engineering, and a well-known TV personality with Forbes Magazine and the Weather Channel*

"Professor Cerveny leads as the world's go-to source for the real story on weather extremes. Here, he gives a fascinating behind-the-scenes look at this important science."

**Richard Alley** *is a named Professor of Geosciences at Penn State University and a member of the US National Academy of Sciences and the Royal Society. He is the author of an award-winning book on climate change and ice cores,* The Two-Mile Time Machine

"*Judging Extreme Weather: Climate Science in Action* is entertaining, informative, and authoritative. Professor Cerveny discusses subjects that are very important today and will become even more important in the future."

**Elbert (Joe) Friday** *is a former Director of the US Weather Service, a retired US Air Force Colonel and a past president of the American Meteorological Society, and a US Permanent Representative to the WMO*

"Everyone has a weather story, but as the 'world custodian of extreme weather records,' professor Cerveny has THE quintessential compilation of wild weather stories. Not only are they fascinating, but by delving into the science of extreme weather, he helps to shed light on the differences between climate and weather… A distinction that is of critical importance for us to understand as we continue to navigate the uncharted waters of the increasing climate crisis. From changes in wildfire and hurricane patterns to extreme heat waves, our planet is undoubtedly changing right before our eyes. Professor Cerveny takes us on a wild ride, investigating the science behind the world's most extreme weather records in a way that's understandable, shocking, and even, dare I say, humorous at times."

**George Kourounis,** *Storm Chaser & Royal Canadian Geographical Society Explorer-In-Residence*

"Professor Cerveny's new book, following the erstwhile TV detective Joe Friday, gives 'just the facts, ma'am!' on worldwide extreme weather. In an entertaining fashion, he draws the reader into climate science by including the interesting 'back stories' of major atmospheric scientists. Fundamentally, this book runs the gauntlet of extreme weather episodes, and will be a major climate resource for years to come."

**Joe Golden** *is a retired Senior Meteorologist at the NOAA Forecast Systems Laboratory in Boulder, Colorado, USA, and is recognized as one of the top international experts on tornadoes and waterspouts*

# Judging Extreme Weather

## Climate Science in Action

Randy Cerveny

Routledge
Taylor & Francis Group
LONDON AND NEW YORK

earthscan
from Routledge

Designed cover image: © Getty Images

First published 2024
by Routledge
4 Park Square, Milton Park, Abingdon, Oxon OX14 4RN

and by Routledge
605 Third Avenue, New York, NY 10158

*Routledge is an imprint of the Taylor & Francis Group, an informa business*



*British Library Cataloguing-in-Publication Data*
A catalogue record for this book is available from the British Library

ISBN: 978-1-032-43570-1 (hbk)
ISBN: 978-1-032-43572-5 (pbk)
ISBN: 978-1-003-36795-6 (ebk)

DOI: 10.4324/9781003367956

Typeset in Sabon
by Taylor & Francis Books

To all of the hard-working women and men of the atmospheric sciences and, in particular, to the two individuals who were instrumental in guiding me into this fascinating, everchanging field, Dr. Robert Balling, Jr. and Dr. Merlin Lawson.

# Contents

# Figures

# 1 Extreme Weather, Who Cares?

> We may achieve climate but weather is thrust upon us.
>
> O. Henry (William Sydney Porter)

In today's world, the phrases *climate change* and *global warming* are often used in political discussions conveyed on the news, on television and within social media. And, yes, let's state this at the very beginning so there are no questions on where I stand. As a climatologist with over forty years of experience studying the Earth's weather and climate, I will state, without a doubt, that our world's climate *is changing*. Of critical importance is the fact that in recent decades it has been changing at a pace that hasn't been seen in millennia—yes, *thousands* of years! That rapid change is caused, to an overwhelming extent, by us, by humanity. I'll hold off discussing the important ramifications of that critical climate assertion until a bit later.

But a vital aspect of that undeniable climate change is often overlooked in those social and political discussions.

*How do we know that it is changing?* In other words, what techniques do we use to exactly measure climate to know for sure that it is shifting? It is like asking a master chef, 'How do you know if you've exceeded the recipe unless you know precisely how much you've already added?'

My inspiration for asking such a simple question dates from the landmark hurricane season of 2005 and, in particular, the gruesome consequences of the storm named Hurricane Katrina. During Hurricane Katrina's landfall in the Gulf States, I was fixated, like most of the country, on the national television news broadcasts of the unfolding disaster in Louisiana as floodwaters inundated New Orleans and the surrounding countryside.

At one point, I took particular notice when a news commentator labeled Katrina as the 'worst hurricane of all time.' Without question, Hurricane Katrina was a horrific storm. I watched the unfolding scenes of death and destruction with profound sadness. Yes, I understand that particularly during times of great stress and trauma, people will often speak in unwarranted superlatives to stress the scale of the disaster.

But that statement—'the worst of all time'—was foolish and misleading. While Hurricane Katrina was a dreadful tragedy with over 1,800 killed, it was not close to being the worst hurricane of all time.

DOI: 10.4324/9781003367956-1

*Figure 1.1* Satellite image of 2005's Hurricane Katrina
Source: Image courtesy of the MODIS Rapid Response Team at Goddard Space Flight Center. NASA Earth Observatory, https://earthobservatory.nasa.gov/images/15395/hurricane-katrina.

In fact, historical research shows that the deadliest hurricane to hit the United States was the catastrophic Galveston hurricane of 1900. It is that tropical weather juggernaut that was made famous by the book *Isaac's Storm*, penned by noted author and journalist Erik Larson. In his book, Larson chronicles the events surrounding that 9 September 1900 hurricane that directly killed *six thousand people* in Galveston, Texas, and another six thousand indirectly, making it—without question—*the deadliest natural disaster* in the history of the United States. Larson's excellent book lays the blame for much of that unfathomable tragedy on the poor state of American weather science at the time and, in particular, to the actions of the Galveston meteorologist, Isaac Cline.

What was the fundamental cause of so much death from a single hurricane?

At the time, the average elevation of Galveston Island was five feet above sea level and, back in 1900, there were no storm barriers—no towering sea walls—existing on the island. Given those two facts, that horrific nineteenth-century monster storm produced a storm surge—the titanic wave-created flooding of a hurricane—that was more than *twelve feet* (3.7 meters) above sea level! In essence, the entire prosperous island was deluged with floodwaters. So many people died that, after the storm, officials were forced to dig mass graves, where hundreds of the storm victims were buried. While six thousand people died due to the storm, many more were killed as law and order broke down on the island. Following the hurricane, a shoot-to-kill order was issued across the island to prevent looting—shoot on sight, no arrests, trials or juries.

As horrific as the aftermath of the 1900 Galveston hurricane was, even that storm doesn't rank as the worst tropical cyclone in recorded history.

Wait, what do I mean by 'tropical cyclone'? Aren't they all called hurricanes? Not quite. Many hurricane-like storms outside of the Atlantic/Eastern Mexico are given different names, such as *typhoon* (for the Asian coast/Pacific), *cyclone* (for India), or, most general of all, *tropical cyclone*. But regardless of their specific regional name, all such tropical storms around the world are of the same nature as hurricanes. They are just called by different regional names.

If we expand our search for a hurricane-like extreme for the entire world, then we find that the worst *tropical cyclone* ever recorded was far deadlier than Hurricane Katrina or even the Galveston Hurricane of 1900. In 1970, a ghastly tropical cyclone struck the country then called East Pakistan (now called Bangladesh). When rescue teams reached the devastated area afterwards, they discovered an unbelievable tragedy had occurred. The 1970 East Pakistan tropical cyclone had killed at least *three hundred thousand people*. Nearly a third of a million people perished in that single hurricane-like storm!

It was a horrific disaster. Like its Galveston counterpart, that tropical cyclone's death toll was generated by a massive storm surge of catastrophic flooding. The storm hit the area of present-day Bangladesh, located in the vast, low-lying marshes of the Ganges River delta, with a storm surge estimated to be over twenty feet (six meters) high. The entire region was inundated with hundreds of people floating out to sea never to be seen again. The cyclone's deadly aftermath even led to a revolution against the government and the creation of the country of Bangladesh.

Therefore, that ghastly storm of 1970 has rightfully earned the appellation of the worst tropical cyclone ever recorded. And, don't worry, I'll talk more about that monster of storm a bit later in chapter 13.

But that brings me back to that news commentator in 2005 who exaggerated the horrific nature of Hurricane Katrina by calling it the worst hurricane of all time. Where were the checks and balances for making such a comment? Who could correct the commentator's misconception?

I decided to find an answer. After searching for such global authorities, I discovered that there wasn't an official listing of the world's weather extremes.

That wasn't the case for individual countries. Many individual countries have created their own specific ways of assessing extremes within their own borders. For instance, since 1997 the United States has had its own means of evaluating weather extremes. The government established the National Climate Extremes Committee (or, for the acronym-lovers, the NCEC), to (in their words) 'assess the scientific merit of extreme meteorological/climatological events and provide a recommendation to NOAA [the National Oceanic and Atmospheric Administration, the nation's umbrella weather and climate management organization] regarding the validity of related meteorological measurements' (National Oceanic and Atmospheric Administration 2023).

I do love governmental technobabble.

In short, the US National Climate Extremes Committee's mission is to judge whether a given weather record in the United States is real and legitimate.

The idea worked. The NCEC is designed around three primary groups of weather experts. First, on the committee, there must be a representative from the nation's primary climate facility, what is called the National Centers for Environmental Information whose central location is in Asheville, North Carolina. Currently, that position is filled by my talented former doctoral student and now learned colleague, Dr. Russell Vose.

Second, the committee must also include members from the National Weather Service, from the specific local forecast office in charge of the area where the extreme occurred. Third, the committee needs to have a member from the American Association of State Climatologists. What is that group? Well, each state in the country has a designated State Climatologist, a person whose primary responsibility is to verify that the weather observations taken in that particular state were properly measured and recorded. One of those people—as an authority on climate records—must be on the committee.

When a potential weather record occurs in the United States, the National Extremes Committee is called together to examine the available evidence for the observation.

For example, back in 2018, the NCEC assessed whether an observation of a record twenty-four-hour rainfall in Hawaii was valid. That rainfall occurred in Kauai, Hawaii where purportedly an enormous rainfall of 49.69 inches (1.26 meters) fell in a twenty-four-hour period during 14–15 April 2018. As part of its investigation, the committee assessed the type of instrument (in this case, a 'tipping bucket' rain gauge, and, yes, more on those later), the regional weather of that day (a large low-pressure system was impacting the Hawaiian Islands), and the observational practices of the weather station (in essence, addressing if the measurement was made correctly). After much discussion, the committee concluded that the observation of 49.69 inches of rain did indeed fall—and that rainfall was properly measured—at Kauai, Hawaii. The NCEC accepted the record as valid.

But not all records examined by the NCEC are approved so systematically.

In early 1999, a location in New York state purportedly recorded a snowfall of 77 inches (1.96 meters) within a twenty-four-hour period. In that instance, the NCEC rejected the observation. Why? Well, oddly enough, they judged that the snowfall observer had made *too frequent* observations! Observations made too frequently? Yes, observing and evaluating the weather can be weird! I'll explore that case a bit later in chapter 9.

Since its creation in 1997, the US National Climate Extremes Committee had accomplished incredible work. But it *only* evaluates extreme weather records for the United States. A few other countries, such as the United Kingdom and Australia, also have created similar type committees to judge the weather extremes that occur within their specific borders.

What about weather records for the whole planet, for the entire Earth?

In my search, I discovered that there were—and still are—a few private organizations such as the *Guinness Book of World Records* that do publish 'accepted' global weather extremes. Back in 2006, I also located a few books that addressed weather extremes, but they tended to focus on sensational weather like *The Elements Rage* by Frank W. Lane, or the legendary Canadian meteorologist David Phillips and colleagues' book *Blame it on the Weather*, or my own book of odd weather anecdotes *Freaks of the Storm*. Finally, I discovered a few excellent weather compilation books written by devoted weather historians such as my good friends Chris Burt, the late (and lamented) Philip Eden and Tye Parzybok.

But none of these sources were 'official'—that is, they weren't accepted and verified by meteorologists around the world. And, unfortunately, my search failed to uncover anything that was actually official for the entire world.

Feeling my oats about the matter (I was a young and naïve professor at the time), I took that absence as a call for action.

First, I contacted three of my colleagues, Jay Lawrimore, Roger Edwards and Chris Landsea. At that time, Jay Lawrimore was the person who chaired the US Climate Extremes Committee, that important group in charge of verifying US weather extremes. Roger Edwards was (and still is) a superb forecaster at the US Severe Storms Prediction Center and a complier of severe weather extremes, with emphasis on tornadoes. I'll profile this superb meteorologist in chapter 11. Dr. Chris Landsea was (and still is) a fantastic atmospheric scientist at the US National Hurricane Center and one of the world's foremost authorities on tropical cyclone extremes, and I'll tell you more about him in chapter 12.

In a series of fascinating discussions, we talked about weather extremes and the need for an authoritative compilation.

We began collecting, evaluating and discussing 'official'—governmental—records of extremes for the world.

Some of those documents were old reports generated by various national agencies. For example, one useful, ninety-four-page document generated by the US Army Corps of Engineers in 1997, was entitled *Weather and Climate Extremes*. It listed many global and continental weather extremes. Why was it written? According to its authors, Paul Krause and Kathleen Flood, the document was designed to 'assist designers of military equipment with information about the extremes of the natural environment' (Krause and Flood 1997, v).

We also uncovered other reports published in the world's major scientific technical journals. For example, one of the earliest notices of a purported hottest recorded temperature for the planet was published all the way back in 1930, in the respected scientific journal *Nature*.

From that diverse information and more, the four of us began to compile a list of government- and science-accepted weather extremes for such variables as temperature, snow, rain and so on. These extremes included records like global highest temperature, greatest twenty-four-hour temperature change, highest twenty-four-hour rainfall, strongest recorded wind and so on. To that (thanks primarily to Roger), we added records for tornado extremes such as the world's deadliest single tornado, the widest recorded tornado, the highest recorded wind speed (estimated from Doppler Radar) and others. And, finally, we included (thanks primarily to Chris) records involving tropical cyclones such as the most intense (as recorded by central pressure), the highest storm surge, the largest- and smallest-diameter tropical cyclones and others.

We saw this work as an opportunity to create a first comprehensive listing of the best-known and accepted weather and climate extremes for Earth to be used by climatologists around the world.

For those global climatologists to use that listing, we needed to publish it. In the science world, that means getting an article accepted in a technical journal, one oriented specifically for scientists. Consequently, in 2006 we submitted an article containing our weather extremes listing for the professional journal *Bulletin of the American Meteorological Society.* The American Meteorological Society is the national umbrella organization of all professional meteorologists around the country—those individuals involved in, for example, the National Weather Service, as well as those who work for private forecasting groups, the military, or in academics.

To my delight after peer review (a means of independent scientific verification that all science articles must undergo), the article was published and well-received by our colleagues around the country—and, indeed, around the world. But it was one passage in the article that began to be noticed by scientists around the world. We had written that 'it would be useful to have the assemblage of existing weather-extremes records mentioned in governmental documentary or online sources compiled into a single updateable official source. . . .' (Cerveny, Lawrimore, Edwards, and Landsea 2007, 859).

Here is where I found that a person must be careful of what they wish for.

Shortly after the publication of that technical article, I received a phone call from Dr. Thomas Peterson, who was a leading figure in a group called the *World Meteorological Organization.* I will have much to say about the extraordinary Dr. Peterson and the WMO in the next chapter.

And with that phone call, I found that my world was about to change. A global 'weather judge' was about to be born.

Before I start discussing my work with the WMO, I need to address one important question. It is a question that I often get asked by the media and the general public: 'Why do we need an official world archive of weather extremes anyway?'

Remember that, when the US Army Corps of Engineers put together their listing, they stated that their weather extremes database was to 'assist designers of military equipment with information about the extremes of the natural environment.'

Undoubtedly, that is an important concern. But are there more socially redeeming (or, at least, less battle-oriented) reasons for creating such a weather extremes archive?

I discovered that there are at least six major reasons for the establishment of such a database.

First is the politically explosive idea of rapid climate change. Wait a minute. How can just measuring and verifying *current* weather extremes be linked to *future* climate change? The answer is straightforward. We need to know—and be confident with—our *existing* weather and climate extremes to determine how much and how fast our world's climates are changing. In other words, a good understanding of our current extremes establishes the critical *baselines* that we need to access how our climate is changing.

Change is generally measured in relation to an accepted value, a baseline.

For example, if you are weighing yourself to determine if you are getting fatter or thinner, you likely compare your present weight to a past accepted weight.

In a similar fashion, common graphs that demonstrate global climate change most often display temperatures in terms of some established baseline, a past reference that scientists accept. Commonly, that baseline is given as the average temperature over a thirty-year period, something climatologists call a 'normal' (and I'll discuss those in more detail in chapter 7).

Baselines are important because they allow us to give context to any weather change, to quantify how much change is occurring. Knowing our world's climatic baselines—in temperature, winds, rainfall and so on—is critical to charting how those climate variables will change in the future. And that is true with both weather averages and weather extremes.

Second, a good knowledge of the world's weather and climate extremes is important for medical and engineering concerns. I've already mentioned how one of the early compilations of world weather was created to assist in the engineering design of weapons and other military equipment. But knowledge of weather extremes is so much more useful than that.

Since the WMO created the World's Weather and Climate Archive and put me in charge, many engineers around the world have contacted me about our weather extremes. For example, if a person is constructing a building or bridge, knowing the limits of wind speed—exactly how fast wind speeds can actually reach—is essential for safe configuration of that structure. Indeed, municipal building codes around the world must address maximum wind loads and other meteorological concerns. Understanding the maximum limits that winds can achieve, both globally and for specific regions of the earth can help ensure safe living conditions. And engineers also need to worry about extremes in temperatures, pressures, rainfall and other weather phenomena, when they are designing their structures.

Knowledge of weather extremes is also important to our human health. How hot can our temperatures reach? How cold? Doctors and medical officials tell us

*Figure 1.2* Demonstrating the need for knowledge of weather extremes?
Source: 'Washington, Tacoma. Suspension Bridge Collapses into the Tacoma Narrows,' 7 November 1940, Photograph, Washington DC, Library of Congress, Prints and Photographs Division, https://www.loc.gov/item/2006687436/.

that our bodies can operate within a specified set of environmental conditions. Identification of those specific limits is important.

For example, one of my past doctoral students, Dr. Kimberly Debiasse, penned an interesting dissertation investigating weather's effect on marathon runners, as she studied the idea of *acclimation*, the process of getting used to a particular environment. She examined whether the weather *preceding* the running event played a role in the performance of the athletes on race day. And she found weather did influence the results. If the weather changed dramatically before race day or that lots of runners had arrived from a different climate, those marathoners tended not to do as well as in the races where the weather before the marathon was similar to the race day itself.

Knowledge of the extremes of weather can help doctors and public health officers determine how their populations will react to weather. Over the past years of this project, individuals from those and other disciplines have shown great interest in some of our weather and climate extremes. Indeed, many universities around the world—including my own—have shown interest in creating departments of sustainability, units designated to find solutions to climate change.

Third, our evaluation of world weather and climate extremes can sometimes advance our basic atmospheric sciences. Although I didn't realize it when I first created the WMO extremes archive, the questions that we ask in our extremes evaluations can lead to new advancements in weather and climate. For example—and I'll discuss this in more detail in chapter 12—one of our recent investigations on lightning flash extremes has caused a fundamental meteorological definition of the word *lightning* itself to be rewritten. Yes, the dictionary definition of lightning has changed because of some of our work.

Fourth, as I mentioned earlier, there is a tendency for a few of the media to overhype a news event such as a weather disaster. We need official and accessible records of weather extremes to aid the media in putting weather events in proper perspective. Flashing forward to present day, since the creation of the WMO extremes archive (and, in large part, due to the wonderful work of the WMO Press Office that I'll discuss in chapter 15), I have found that many news outlets around the world now contact the WMO and me *before* publishing a news article regarding potential weather extremes. We are having an impact.

Fifth, and surprisingly to some, there are many locales around the world that commemorate and recognize the occurrence of major weather events. For example, a huge sign at Mt. Washington Observatory in New Hampshire, USA acknowledges their long-held record for the highest human-recorded wind of 231 miles per hour (a record recently exceeded by a powerful cyclone's wind gust at a small island off Australia—and I'll talk about that investigation and the fantastic Mt. Washington Observatory in chapter 6). Other locales have similar recognition of their extremes.

Of course, the converse also holds true. There are some places that *don't* want to be recognized for their severe weather. As you'll see throughout this book, many locations around the world have suffered tremendous pain and human loss due to extreme weather (for example, the horrific events discussed in chapter 13). Such places might not want to recognize their weather-torn past. But I—and other meteorologists—believe that it is critical that such extremes are documented. As the famous saying might be paraphrased, those people who forget their weather extremes could be doomed to suffer even greater loss in the future.

Sixth, people in general are fascinated by weather and they love weather extremes—the hottest, the coldest, the windiest and so on. Having a reliable list of these extremes helps foster people's interest in weather. And even win bets!

Some emails and phone calls that I have received since the creation of the WMO World Archive of Weather and Climate Extremes have indicated that I have helped people win some lucrative bar bets (well, rewarding at least in terms of free beers) about various weather extremes. I must admit that was not something I had anticipated when we created the WMO Extremes Archive back in 2007, but it does show how much interest exists in the public about this type of records.

Related to that public interest is one of the unexpected uses that I've discovered for the list of worldwide weather and climate extremes. Through emails, letters and from their enthusiasm during my presentations, I have discovered that children are fascinated with weather extremes. I have tried to carry that interest further into an interest in meteorology. By grabbing the interest of kids around the world with these weather extremes, we are promoting possible future careers in the atmospheric sciences and ensuring that we will have quality meteorologists and climatologists for the future. And, after nearly two decades of this, I can report with extreme gratitude that, yes, some meteorologists who are working today say they are in the field because I made them excited about it back when they were kids.

Those six reasons address the valid question of 'Why do we need an official world archive of weather extremes anyway?'

But that still leaves the underlying question from our technical paper: Who should be the group in charge of such an archive? What official international agency exists to monitor world weather extremes?

The answer to that question involves a look back at one of the first attempts to achieve international—*global*—cooperation between nations.

An attempt that even predates the United Nations.

A little-known effort in the nineteenth century to unite the world's weather services.

# Interlude: Introducing *Freaks of the Storm*

Okay, I admit it. I love unusual and bizarre weather! That love has inspired some of my work with world weather extremes. But my interest in weird weather does go beyond the serious business of verifying weather extremes. In fact, a few years ago—as my first term as 'world weather judge' was beginning—I penned a humorous little book called *Freaks of the Storm*. I designed it as an amusing collection of bizarre and unusual weather stories. Its strange title originates from what old-timers in the Midwest used to do at their barber shops and other places following a big storm—'chewin' the fat about the freaks of the storm.'

One of the pleasures of being the curator of the world's weather extremes archive is that I am still often consulted whenever something weird or unusual occurs with the weather. That has allowed me to continue to add to my huge database of strange and unusual weather events. And, as I give talks and presentations around the country, I've discovered people do enjoy hearing these odd stories.

So, as fun little sidelights between my more serious chapters, I will indulge my child-like love of weird weather by relaying a few odd and unusual freaks of the storms. If these fun little interludes aren't your particular cup of tea, please feel to move to the next chapter where I will again get serious about weather.

Let me introduce the idea of freaks of the storm with a strange 'miraculous wind' story from Africa that took place in late October of 2021. During one clear Wednesday afternoon, the residents of Ondo Town in Nigeria experienced what they believed to be a weather-related heavenly miracle. As the villagers of the small town watched in awe, a long undulating strip of white fabric—to the residents, obviously divine—literally *fell from the sky*. But, when the wind-blown cloth floated down to earth, the supposed miracle turned violent. The villagers began to fight and wrestle each other to grab a piece of what they assumed to be divinely blessed cloth for themselves. They ripped the long snow-white fabric into complete shreds. As a witness to the melee, Fatima Yesufu, reported, 'I heard some strange voices, came out and saw a lot of people struggling with a big cloth . . . the stampede was so much that I had to stay away so they would not stamp on me' (Nurudeen Lawai 2021).

As is usually the case, the truth turned out to be less miraculous. Officials learned that white fabric belonged to a local resident named Agnes Fadoju, who had purchased the very long bolt of white cloth—a hundred and twenty yards—for her decorating business. After washing the fabric and setting it outside to dry, she said that she began to receive 'lots of phone calls' from friends asking if she had seen what was happening to her cloth. Ms. Fadoju subsequently learned that 'a wind had blown off one of the clothes to the next street in a way that looked mysterious and everything had now been mutilated' (Nurudeen Lawai 2021).

From the video I saw of the event, in my professional judgment, I would say that the bolt of cloth had most likely been picked up by a strong dust devil. Also speaking about the incident, Ahmed Balogun, an applied meteorology professor in the Department of Meteorology and Climate Science, Federal University of Technology, in Akure, Nigeria said, 'It is very possible to have the cloth blown from where it was hung to the sky by the wind . . . we don't normally see [the air as it is a] gas but it is always there and it is part of it that forms the clouds that we see' (Nurudeen Lawai 2021).

Ms. Fadoju took the loss rather philosophically. When she later spoke to an official in the community, she said, 'If I had seen the 'mysterious' manner [in which] the cloth descended, I would have believed [the villagers'] claim' (Nurudeen Lawai 2021). And fortunately for her decorating job, we are told, the event did end up providing significant publicity for her business.

In a similar fashion, residents of the small town of Olten between Zurich and Basel in Switzerland also experienced a strange *falling from the sky*. But here, rather than divine fabric, the stuff snowing down from the sky was a bit tastier. One week in August of 2020, the locals woke up to a bizarre snowfall of *chocolate*!

The incident began on a Tuesday at the nearby Lindt & Spruengli factory, a plant that makes the world-renown Lindt chocolate. A malfunction in the factory's ventilation system along the production line of roasted cocoa nibs caused the cocoa products to be emitted into the surrounding air outside of the plant. Cocoa nibs are tiny powdery fragments of crushed cocoa beans that form the basic ingredients for making chocolate.

During the week, the winds picked up speed so that, by Friday, residents of nearby Olten reported that it was snowing cocoa powder! So much cocoa fell that cars in the area were even lightly coated in chocolate. The company did offer to pay for any cleaning needed. Company officials also were quick to say that the particles were completely harmless to people or the environment. And, yes, the factory's ventilation system was quickly repaired.

# 2 Enter the WMO

The weather is like the government, always in the wrong.

Jerome K. Jerome

I received a phone call from Dr. Thomas Peterson soon after the 2006 publication of our professional meteorology article about weather extremes and the need for a global extremes database. Dr. Peterson at that time was a major expert with a group called the *World Meteorological Organization* or, as always with government groups who love acronyms, the WMO. At that time, I heard of the organization's existence, but didn't know too much of its history.

To my surprise, I discovered that the WMO was formed as one of the world's first attempts at international cooperation. Its beginnings date back to 1872.

Many people believe the modern era of international cooperation between nations only began in the early twentieth century. Global collaboration resulted, they said, after people had seen World War I's horrific destruction. The war's deadly aftermath spurred the desire of many at the time to prevent such catastrophes from occurring in the future. People across the world asked, 'Can we create an organization of the world's various countries that works to ensure world peace?'

Those history scholars often cite the creation of the League of Nations in 1920 as one of the first global attempts to unify humanity and encourage international unity. That group was charged with creating and maintaining international cooperation and peace. Countries established the League of Nations with the goal of preventing the widespread suffering and destruction seen in the Great War. Indeed, it was during the post-war Paris Peace Conference in January 1919 in which the Allied powers agreed on the idea of the League of Nations. The League—an organization, most modern historians state, which was not perfect and suffered many problems—nevertheless remained in place as a start towards global unity until it was disbanded in 1946, following World War II. At that time, its powers and functions were transferred to its heir, the present-day United Nations.

But is it true, as many people believe, that global unity only began with organizations like the League of Nations and the United Nations?

DOI: 10.4324/9781003367956-2

It might be surprising to learn that advanced *scientific* cooperation on a vast international, global scale had commenced fifty years *before* the League of Nations—and that cooperation involved the monitoring of weather.

In 1872, a small group of fifty meteorologists met at a scientific weather meeting held in Leipzig Poland. Wait a moment. Scientists promoting global unity in a conference? Even today, many people imagine that science conventions held around the world are filled with hours of technical mumbo-jumbo and are of little importance in solving 'real world' issues. This conference back in 1872 would prove that misconception to be quite wrong.

During that meeting convened to address urgent scientific weather concerns, those early meteorologists recognized the critical need for full international—*global*—standardization of weather measurement. By that, they meant that weather observations made in Britain should be made in the same way, and using the same comparable equipment, as similar observations conducted in the Netherlands, France, the United States or any other country of the world.

One of those scientists, a renowned meteorologist named Professor Buys Ballot of the Netherlands, stated the group's objectives in simple terms (well, simple, at least, for scientists): 'It is elementary to have a worldwide network of meteorological observations, free exchange of observations between nations and international agreement on standardized observation methods and units in order to be able to compare these [weather] observations.'

His statement was a recognition of the fact that the industrializing world of the mid-1800s was becoming more and more unified by trade and interaction. A growing network of steam and sailing ships and, across land, by locomotives connected distant places. And that unification meant critical environmental information taken at one location was being used by people who lived and worked in far different places around the globe.

Think of a nineteenth-century sailing captain as he was charting a course around the world. Would the weather observations that he received in one port be comparable to those of another? For instance, were the two observations of temperature taken at the same height with the same type of equipment? And were they reported in the same units of measurement? (As I will discuss later on in chapter 7, that problem of 'units', even today, is still a major concern).

Until 1872, there were no assurances that such weather observations made in different locations around the world were measured in similar manners.

This issue was considered important enough to the world's top meteorologists, that they proposed back in 1872 to address the problem in a much bigger meteorological conference to be held the very next year. And, in that 1873 conference, called the Vienna International Meteorological Congress, a group called the 'Permanent Meteorological Committee' was created to draft specific rules and statutes of an international meteorological organization to 'facilitate the exchange of weather information across national borders.'

That 1873 conference's permanent committee issued two proposals that—for that time in particular—we would today consider far ahead of their time. One proposition was for the formation of an International Meteorological

Institution—a worldwide meteorological organization with a paid secretariat. In essence, they proposed a scientific United Nations for weather. The second motion proposed the establishment of an International Meteorological Fund—a monetary account to be used to create remote weather observatories across the earth's surface. These would be elaborate *manned* weather stations as there was no automated equipment back at that time.

Although the committee endorsed both of these proposals in principle, the meteorologists were realistic enough to consider their propositions were unlikely to be implemented because of the enormous administrative and financial concerns. But their work sowed the seeds for global cooperation in weather back in those early days. That one science conference, held one hundred fifty years ago, is now recognized as the start of the organization called the World Meteorological Organization.

One primary need that those meteorologists discovered was the necessity of linking any *planetary* network of weather observations to the already-existing *national* weather networks around the world. That meant that the leaders—the chiefs—of those many individual weather organizations needed to be invested in this proposed global ideal of weather cooperation. By the 1880s, a new International Meteorological Committee (the successor to the 1870s permanent meteorological committee) was formed with all nine of its members being the directors of their respective nations' weather services. That committee included members from countries such as Russia, the Netherlands, France and Britain.

Okay, I realize that these many committees are starting to merge together. But it was that new 1880s government-oriented committee that began to promote international cooperation in meteorology. It encouraged meteorological research and to establish uniformity in operational practices. Its members paid special attention to the standardization of weather reporting and to the exchange and publication of weather data across the entire world. In particular, the committee proposed all countries should implement common methods of weather observation. It urged that nations should standardize their weather instruments and those countries should publish all their weather observations.

That committee's work paid off. For example, one of the first comprehensive sets of truly international Meteorological Tables—global listings of temperatures, pressure, rainfalls and other data—was published in 1889. Additionally, meteorological stations were established in many countries in important, but remote, regions of the world, places where observations had hitherto not been systematically conducted.

By 1889 still another meeting of meteorologists, the International Meteorological Congress held in Rome, formally created the International Meteorological Organization (or the IMO). And, in 1891, the IMO expanded its membership so that the conference attendees were identified first as *meteorologists*, not as representatives of governments. And this IMO, as a nongovernmental organization, served the causes of international meteorology until the start of World War I. Up to that time, governments and meteorologists alike were well satisfied with the work of the IMO.

And then global war broke out.

The Great War proved that specific knowledge of meteorology could help—or conversely even destroy—military forces. For example, during World War I, while on a bombing raid on London, planes of the German Air Force were caught unprepared. An unexpected powerful northeast wind blew the planes towards French lines whereupon the aircraft were shot to pieces by antiaircraft fire. Military leaders began to appreciate that, if those types of winds could have been forecast, the dire and often deadly situations such as experienced by the unfortunate German pilots could be anticipated—and prevented.

As the war continued, countries began to appreciate that weather observations could be considered valuable war information. And that realization, unfortunately, caused global cooperation regarding the exchange and standardization of weather observations to diminish.

Such weather secrecy was accelerated as our basic knowledge of atmospheric science grew. As an illustration of the critical importance of the *science* of meteorology to the war effort, let me mention the work of a brilliant meteorologist named Vilhelm Bjerknes. Bjerknes oversaw what today we would label a think tank of Norwegian scientists. This was an incredible idea.

Imagine an early-1900s Manhattan-style project but rather than developing atomic bombs, Bjerknes' group focused on better understanding the nature of weather itself. From their work, that Norwegian group of meteorologists developed a major theory that elegantly explained how weather, and in particular a storm, moves and changes across places like North America and Europe. The conceptual model that they developed was called 'polar front theory.' It is still used today and still incorporates some of the military terminology from World War I.

Bjerknes and his colleagues envisioned weather as the result of huge battling air masses—like aerial armies of gases—each of a different temperature and humidity. The battleground of these air masses, the place where storms were born, were called 'fronts' after the battlefields between armies common in World War I. The character of the 'winning' air mass—the aerial army that was advancing—determined the name of the front (e.g., 'warm front', 'cold front'). That militaristic naming system—and, indeed, the weather theory itself—is still an integral part of meteorology today.

Bjerknes' development of polar front theory was a game changer in warfare. For the first time in history, generals could receive accurate forecasts about future battle conditions—and plan accordingly. After World War I, international cooperation in meteorology again grew, particularly as aviation advanced. Accurate and detailed weather information was critical to the pilots of the planes and airships that had begun to fly the skies.

But it was that advancement of aviation that led to the next stumbling block for international weather cooperation. By the start of World War II, nations once again began to hoard their weather information. Our developing attempt at global unity, the International Meteorological Organization, was functioning but it had been reduced to a mere shell of its former self, as war waged across the globe.

The military's demand—and hoarding—of weather knowledge is illustrated in the secretive discovery of jet streams. Today, we know that jet streams are narrow, intense 'rivers' of fast-moving air that blow from west to east. They are found high in the atmosphere at a height of thirty-five thousand feet (10.7 kilometers). Most people of modern times have experienced the effects of jet streams when they have flown across North America, over the Atlantic or across Europe. Due to jet streams, air travel from west to east, say from Los Angeles to New York, can be much faster than the reverse because the jet streams help to push the eastbound planes with a strong tailwind, sometimes well over a hundred miles per hour. Conversely, westbound planes often must contend with strong headwinds as they fly against a jet stream, slowing them down.

For example, planes from Los Angeles might arrive at New York thirty minutes ahead schedule, while the same flights from New York to L.A. might land thirty minutes behind schedule. All of this is because of the jet stream.

However, our discovery of jet streams only dates to the 1930s and the eventual commencement of World War II.

At that time, a German meteorologist named Heinrich Seilkopf was one of first to observe and to theorize about these high-level wind rivers. He actually gave them the name that we use today, 'jet streams.' At first no one else appreciated the importance of his discovery—and, perhaps under orders, he kept mum about them.

This was because in the 1930s Seilkopf was entrusted with the responsibility of plotting the courses for the great zeppelin airship fleet. At that time people were amazed that the German zeppelins, hydrogen-filled airships that were only equipped with relatively weak steering motors, were still able to accomplish many of their flights not only on time but often in record time. To their competitors' bewilderment, the zeppelins frequently arrived at their destinations far faster than the weak zeppelin motors should have permitted.

*Figure 2.1* A 1910's Zepplin passenger ship (associated with the secretive 'jet stream')
Source: Image courtesy of the Library of Congress, Call Number: LC-B2- 2241-13 [P&P], Repository: Library of Congress Prints and Photographs Division Washington, D.C. 20540 USA http://hdl.loc.gov/loc.pnp/pp.print

People at that time attributed the zeppelins' legendary travel speeds to the Germans' vaulted skill in air flight. But it would have been more appropriate if those people had admired the brilliance of German meteorology—and the secretive discovery of jet streams. What the Germans had learned was that if they lifted a zeppelin high up into a jet stream, they could use the jet's tailwind advantage to fly at unparalleled speeds to the airship's destination.

Whether right or wrong, that information remained proprietary with the Germans. Indeed, the usefulness of jet streams was not appreciated by the English until World War II. In 1943, a fleet of English bombers ran out of fuel over occupied France as, bewildered, they battled against a jet stream's incredible headwind of over 120 mph (193 kph). A year later in late 1944, the Americans found themselves in a similar situation when their westward-flying B-29 bombers slowed to a crawl as they flew into extreme winds over Japan. Unknown to the Americans, they had encountered, like the English a year earlier, a very strong winter jet stream.

In essence, World War II reemphasized what had been learned in the First World War: that, in some circumstances, weather knowledge could—and in the minds of many, should—be considered military state secrets.

That nationalistic attitude lessened after the war in the late 1940s. The old International Meteorological Organization resumed its work of promoting international weather cooperation.

A new cooperative meteorological spirit was evident in the agenda for the IMO meeting that convened in September 1947. Over four hundred resolutions were introduced in that hardworking meeting covering an incredible range of weather topics. Weather delegates proposed international guidelines on standardizing weather codes and units, on regulating instruments and methods of observation, on proper construction of station networks involving telecommunications, and on the safety of air navigation.

They introduced more resolutions on the need for comprehensive climatological statistics, the need for weather documentation, as well as improved education and professional training, meteorological research, legal issues and administrative matters. Moreover, the IMO began to branch out, proposing and initiating cooperative relationships with several new groups, such as International Civil Aviation Organization, the International Telecommunication Union and the International Ice Patrol.

But the issue of greatest importance to the IMO after the war was the organization's own status and structure.

The two major global wars in the first half of the twentieth century had proven that meteorologists and governments *needed* to be working together. With that in mind, the IMO directors prepared a new constitution, which would make the organization an intergovernmental body linked to the various nations' weather services. Then, in 1950, IMO rebranded itself as the World Meteorological Organization (WMO).

A year later, the WMO formally joined the United Nations as a 'specialized agency' of that organization. As such, the WMO is tasked with creating and maintaining international cooperation and coordination concerning all aspects of Earth's atmosphere, its interaction with the land and oceans, the weather and climate it produces, and of the Earth's water resources.

This shows, since its origins date back to 1872, the World Meteorological Organization has played a unique and powerful role in contributing to the safety and welfare of humanity.

Oddly enough, even though the WMO does accomplish many good things and has such a storied and interesting past, most people across the globe haven't heard of it.

With one exception.

During every tropical cyclone season, one particular responsibility of the WMO does touch almost every person on the planet. Since 1953, every hurricane that has affected the United States (and, indeed, tropical storms impacting the rest of the world) has been named by the World Meteorological Organization. The WMO maintains a rotating set of names deemed appropriate for each ocean basin around the world in which tropical cyclones can occur. That means all major hurricanes like Katrina, Andrew or Sandy were so named by committees of international meteorologists working with the WMO. Why do they do that? I'll save that discussion for a bit later when I discuss tropical cyclone extremes in chapter 10.

Beyond that, the WMO does (at least in my admittedly biased opinion) incredible work. As a taste of its wide diversity, some of the WMO's recent activities have included:

- the establishment of an integrated global 'Earth System' observation network to provide weather, climate and water-related data,
- the establishment and maintenance of data-management centers and telecommunication systems across the world for the provision and rapid exchange of weather, climate and water-related data, and,
- most importantly to my own work, the 'creation of standards for observation and monitoring in order to ensure adequate uniformity in the practices and procedures employed worldwide and, thereby, ascertain the homogeneity of data and statistics.'

The WMO has also branched out to work with weather, climate and water-related services to reduce disaster risks and contribute to climate change adaptation, as well as for sectors such as transport (aviation, maritime and land-based), water resource management, agriculture, health, energy and other areas. For example, I am pleased to say that one of our investigations on lightning extremes (discussed in chapter 12) has helped in the creation and promotion of an International Lightning Safety Day held every June.

In terms of the important issue of climate change, one creation of the WMO has achieved major global attention over the past few decades. In partnership with the United Nations Environment Programme (UNEP) in 1988, the WMO formed the Intergovernmental Panel on Climate Change (IPCC).

The primary goal of the IPCC has been to provide nations across the world with scientific information that they can use to develop climate policies. The famous IPCC Reports—there have been six since the Panel's creation—have been key inputs into international climate change negotiations. Following the WMO's current design, the IPCC is an organization of governments that are members of the United Nations and/or the WMO.

Okay, history has demonstrated the WMO has been—and is—an impressive *global* organization dedicated to understanding weather and relaying that knowledge to the world.

Remember that, when my colleagues and I wrote that paper back in 2006 on world weather extremes, we had written that 'it would be useful to have the assemblage of existing weather-extremes records mentioned in governmental documentary or online sources compiled into a single updateable official source. . .' (Cerveny, Lawrimore, Edwards, and Landsea 2007, 859).

The World Meteorological Organization was the logical place for our proposed Archive of World Weather and Climate Extremes. One of its members, Dr. Thomas Peterson, suggested that I give a presentation of my idea to the WMO. That request made me nervous.

Tom Peterson, now retired, has long been one of the most important gurus in the field of climatology. At the time he called me, he was a top scientist at the

*Figure 2.2* Left, the World Meteorological Organization Headquarters Building in Geneva Switzerland. Right, Dr. Thomas Peterson, noted climatologist and past president of the Commission for Climatology, WMO

Source: Photographs courtesy of Peterson

National Climatic Data Center (now called the National Centers for Environmental Information). He was a lead author of one of those IPCC Reports I mentioned above, and (back in 2006) he would soon be the co-editor-in-chief and cochair of one of the United States Global Change Research Program's major reports on global climate change impacts. Already, by 2004, he had been named in the top one percent of most cited researchers in the entire field of geoscience. A literal powerhouse in the field of climate.

This climate superstar was asking me to present my little idea at an international meeting?

In 2006, I travelled anxiously—yes, I even initially boarded the wrong plane!—to a WMO meeting in the beautiful picturesque town of Tarragona Spain (near Barcelona). And there I presented my vision of a world archive of weather and climate extremes to a committee of distinguished international meteorologists. This particular meeting included some of the highest-regarded atmospheric scientists in the entire world!

Somehow, I managed to stumble through my presentation without major embarrassment.

And then I waited as they began to debate the merits of my idea. Would I be laughed out of the meeting hall?

No! They were enthusiastically supportive. That group of WMO atmospheric scientists thought that the creation of a global archive of weather and climate extremes would be useful. And so, they unanimously voted to carry out the idea—with one major change from my original design.

Would I *personally* be willing to lead an effort to create a global archive of extremes?

Not realizing what I was getting myself into, I agreed.

And so, I was named the WMO's first Rapporteur of Weather and Climate Extremes.

Wait, *rapporteur*? What is that?

The impressive name referred to what I had presented at that meeting in Tarragona Spain.

According to the Merriam-Webster Dictionary, a *rapporteur* is 'a person who gives reports (as at a meeting of a learned society)' (Merriam-Webster, 2023). The word was first adopted into English back in the sixteenth century and is a descendant of the Middle French verb *rapporter*, meaning 'to bring back, report, or refer.' It is now used by organizations such as the United Nations to denote a specialist in a particular area of interest and one who oversees aspects of that area of interest for the organization.

That starlit evening, I gazed down at the impressive set of illuminated ancient Roman ruins in Tarragona and rubbed my hands together anxiously. The WMO had named me a Rapporteur in charge of creating a WMO Archive of World Weather and Climate Extremes.

Now the work was to begin.

# Interlude: Freaks of Political Weather

The World Meteorological Organization, as part of the United Nations, is a political, as well as scientific, group. Therefore, in this little interlude, I will focus on a few of the odder politically related weather anecdotes that I have come across. I didn't have much trouble finding such stories. Anecdotes on the political aspects of weather date back to ancient Roman days.

For example, the esteemed Roman politician and lawyer Cicero in a debate with his brother on the issue of national fortune telling through lightning observations, noted wryly, 'We regard lightning on the left hand as a most favorable omen, for everything except an election. No doubt this exception has been made to allow the rulers of the State political expediency to decide the correctness of an election for magistrates, judges or legislators' (Cerveny 1994, 10).

In fact, in ancient Roman times, the mere sight of any lightning was enough to force the cancellation of all public assemblies for the day. By the time of Caesar, shrewd politicians had exploited this rather restrictive statute by employing it as a device to postpone unwanted meetings of the Public Assembly and to cancel the results of popular elections to which senior politicians objected. The phrase 'I will watch the sky' became synonymous with casting a veto against a political action.

In 59 BC, the consul Bibulus delayed the entire legislative agenda of Julius Caesar for a time by requesting that the augurs watch the sky for signs of lightning. The following year, the Romans passed a law to prevent such obstructive tactics from being employed against legislation passage.

Throughout history, political debates involving weather have become quite partisan. Politics even invaded the discussion of important meteorology technology. For instance, the inventor of the lightning rod, Benjamin Franklin, was convinced that only pointed metal rods made good lightning conductors. Others held to the view (which has since turned out to be correct) that rounded lightning conductors were the best. One of the staunchest upholders of lightning rod knobs was no less than King George III who tried hard to win Sir John Pringle over to his point of view. Sir John's reply was a masterpiece of political tack and conviction: 'Sire,' he responded, 'I cannot reverse the laws and operations of nature' (Anonymous 1856, 470).

Of interest to viewers of today's political environmental discussions, the lightning knob versus point debate became a touchstone for political affiliation in Britain and the colonies. Men who advocated sharp lightning conductors were classified with Franklin and the revolutionary party, while those who favored blunt conductors were held to be loyal subjects of the king.

Some leaders have been big supporters of meteorologists—others not so much.

Back in 1863, President Abraham Lincoln proved that he wasn't always a fan of meteorology or at least some practitioners of the science. After reviewing the application of one Francis L. Capen for a weather forecasting job, Lincoln wrote, 'It seems to me that Mr. Capan knows nothing about the weather, in advance. He told me three days ago that it would not rain again till the 30th of April or the 1st of May. It is raining now and has been for 10 hours. I cannot spare any more time to Mr. Capen.'

Meteorologists can sometimes get the upper hand against politicians. One rather ingenious meteorologist derived a novel means of attacking 'antiweather' politicians. Back in 1887, Clement L. Wragge of Australia became one of the first scientists to name tropical cyclones and other weather features.

Wragge's weather charts showed the daily location of weather systems over Australia and the surrounding oceans. Rather than christening each identifiable system of high or low after girlfriends or the Greek alphabet (as some later researchers would do), in a few cases Wragge paid off grudges by naming the nastiest storms after his enemies. For example, Wragge named three storms after three Australian politicians, Drake, Barton and Deakin, who had voted against additional appropriations to the country's Weather Bureau. Of course, his enemies retaliated by referring to the forecaster as 'inclement Wragge' or 'Wet Wragge.'

To end this interlude, let's look at an instance where forecasters and politicians were working on the same page—to save lives! At the beginning of the Spanish-American War in 1898, then head of the US Weather Bureau, Willis L. Moore, had coerced a meeting with President William McKinley to discuss an urgent need. In that meeting, he stated that far more US ships had been sunk by weather than by war. Moore urged that, to counter that horrible fact, a weather warning service for the West Indies should be immediately implemented.

According to Moore's recollection of the meeting, 'I can see [President McKinley] now as he stood with one leg carelessly thrown across his desk, chin in hand and elbow on knee, studying the map I had spread before him. Suddenly he turned to the Secretary [of Agriculture and Moore's direct boss] and said: "Wilson, I am more afraid of a West Indian hurricane than I am of the entire Spanish Navy." To me, [the President] said: "Get this service inaugurated at the earliest possible moment"' (Cerveny 2019, 35).

# 3 Operational Parameters, Choosing the Best of the Best

> Meteorology is the science concerned with the Earth's atmosphere and its physical processes. A meteorologist is a physical scientist who observes, studies, or forecasts the weather.
>
> US National Weather Service

After that stressful (for me) meeting in 2006 in Tarragona Spain, I had permission to set up the WMO's World Weather and Climate Extremes Archive. One big question now loomed before me: How would the Archive actually work?

That was the question I faced when I returned to the hot desert confines of Arizona State University in the American Southwest where I have now worked for almost forty years. We were creating the Archive from the ground up. I barely had enough time for my talented team of web designers to create the website in 2007 when I was informed of a potential new weather extreme, a record rainfall in Mexico.

At that time back in the Archive's early days, the US National Hurricane Center released a review report entitled *Tropical Cyclone Report: Hurricane Wilma* by Richard Pascha and colleagues. That report stated a massive amount of rain had fallen over a Mexican weather station on an island off the Yucatan Peninsula during the hurricane's passage. It reported that, over the course of a single day in October 2005, an astounding 1633.98 millimeters, more than *sixty-four inches*, of rainfall fell—as recorded by an automated weather station—on an island off the Yucatan. The cause of this enormous rainfall was the passage of catastrophic Category 4 Hurricane Wilma over Isla Mujeres, an island located just off the eastern coast of Mexico near Cancun.

I began an initial search for information. First, I determined that rainfall amount—over 1.6 meters of water—was still *below* the twenty-four-hour rainfall record for the *world* that we accepted at that time (and which is still the record for a twenty-four-hour rainfall). That world-record amount is a mind-boggling 1.825 meters of rain—seventy-two inches, in English units—recorded over a twenty-four-hour period on the small French volcanic island of La Reunion in the South Indian Ocean associated with the passage of a tropical cyclone.

DOI: 10.4324/9781003367956-3

*Figure 3.1* Hurricane Wilma as imaged by the Moderate Resolution Imaging Spectroradiometer (MODIS) on NASA's Aqua satellite on 20 October 2005.
Source: NASA image courtesy Jeff Schmaltz, MODIS Rapid Response Team, Goddard Space Flight Center, https://www.nasa.gov/vision/earth/lookingatearth/h2005_wilma.html.

But I couldn't find any *Western Hemisphere* record of a higher precipitation extreme than the 2005 Mexican value from Hurricane Wilma. Could this value be a contender for the highest Western Hemispheric twenty-four-hour rainfall extreme? It seemed to me to be a good chance to put the WMO World Archive of Weather and Climate Extremes to work.

The initial information given in that US National Hurricane Center report on Hurricane Wilma set some of the standards for how the WMO determines whether to evaluate a record. Generally, we wanted a first mention of a potential record to be from a professional publication or from an official news release (or, often in today's world, a social media post from a weather expert. In particular, I like to have the first mention of a potential record announced by a recognized authority—like a country's National Weather Service.

Why not let anybody submit a record? Well, we do—or rather, we don't limit anybody from suggesting a possible evaluation. But suppose, for example, that on a hot summer afternoon some person—let's call him 'Joe Public'—notices that his car thermometer idling on an asphalt parking lot registers a temperature of 140°F (60°C). Joe then posts or emails me that he has set the world-record highest recorded temperature extreme. Of course, he urges that I should immediately update the WMO website with Joe's new personal information. In general, I don't pay quite as much attention to Joe's proclamation as I would from, say, the head of an Antarctic research station emailing me (or tweeting) that the station has exceeded a temperature record for that continent.

You may say that's not fair, but given limited resources and time, the WMO must restrict the number of 'wild goose chase' evaluations. Frankly, trained meteorologists from a country's weather service are more qualified observers—and have better equipment and recording procedures—than most members of the public. Most importantly, recall that there are very specific world standards for making weather observations—remember the reasons of the WMO's formation back in the 1800s?

With that said, I have initiated examinations based on less-than-full-governmental announcements. For instance, as I will detail in the next chapter, one of our most important re-evaluations, that of the Libyan world-record temperature of 58°C (136.4°F), was initiated on new information supplied by a private climate historian and a heroic Libyan climatologist. But in general, the WMO likes to have official notifications of possible new extremes, ideally from officials of one of its member country's national weather services.

Based on the US National Hurricane Center's Hurricane Wilma report, I contacted Dr. Tom Peterson, the WMO head of the climate division at that time. I outlined the rainfall extreme to him and told him, 'It looks like a good opportunity to start a formal WMO investigation.' He agreed . . . and the fun began.

We first needed to create an official ad-hoc WMO evaluation committee for each potential record. This would be the critical panel of experts who would carefully examine the documentation of the purported extreme observation, discuss it, and then render a reasoned recommendation to me on whether or not to accept it.

How should I form such a committee?

As I mentioned in the first chapter, individual countries like the United States or the United Kingdom have long had their own 'extremes' committees to examine extremes occurring within their borders. The members of those committees are usually pre-set to include members of the local Weather Service, from the country's national level and from associated scientists, like, in the case of the United States, the group of State Climatologists. But we had nothing similar at the global level. We needed to start from the ground up.

From the beginning, I had several goals in mind that I envisioned for the WMO extreme committees that would conduct formal scientific evaluations of weather extremes.

First, I appreciated that the World Meteorological Organization was—and is—a *global* assembly of weather experts. So, in my view, these official ad-hoc WMO evaluation committees should reflect that characteristic—they should be *international*. For example, I didn't want an evaluation committee composed of only Americans, or any other single country. Now, after more than sixteen years of work, I am pleased to note that my committees have included more than 150 members from over forty different countries.

Second, I wanted the findings of the various evaluation committees to be respected by the entire scientific community. That meant the members of those WMO committees needed to be 'blue-ribbon'—in other words, the 'best of the best' in atmospheric sciences. And, yes, in the beginning for an unsophisticated professor from Arizona, contacting those international superstars of meteorology was a bit intimidating. Think of your own choice profession—and how you would react if you found it necessary to contact the top people in *your* field!

But I was determined. And those leaders in atmospheric sciences were friendly to that naïve professor back in 2007—and intrigued with the idea of verifying global weather extremes. I discovered that the superstars of meteorology were available and, most importantly, they were interested in being part of this project.

Now, looking back after more than sixteen years into the project, I have found that our early successes were so notable that I now have some well-known and respected atmospheric scientists directly contacting me. They have told me that they would like to take part in any future evaluations. Success does open doors!

Third, I wanted my WMO committees to be diverse. Unfortunately, in its early past, meteorology—and many of the natural sciences in general—have tended to be white male-dominated. Additionally, as I delved into the WMO's history, the make-up of many early high-profile international committees were older scientists. Nothing wrong with old age per say—I'm now pretty old myself—but it would be nice to have a variety of different thoughts and opinions as we evaluated these records. Many members of the WMO, including my good friend Dr. Tom Peterson (and later his successor Dr. Manola Brunet), and I wanted to do something to ensure diversity in our evaluation committees.

I need to stop a moment and recognize a superb climatologist and good friend.

Dr. Manola Brunet is a powerhouse atmospheric scientist. She was the first woman elected president of the World Meteorological Organization's (WMO) Commission of Climatology (CCI)—one of the top positions at the WMO. Still today, she somehow manages to wear an incredible number of hats—she is a professor in the Department of Geography of the Universitat Rovira i Vigili (URV) in beautiful Tarragona Spain, she is the head of the Climate Change Centre at that University, she is a visiting professor in the famous (at least to atmospheric scientists) Climatic Research Unit at the University of East Anglia (Norwich, UK)—and she is, without question, one of the top people in the WMO. She is one of those rare people who can get things done—and done right.

Dr. Brunet is a renowned expert in the field of instrumental reconstruction and climate analysis. In the course of her research, she has had to overcome many challenges related to the insufficient quality of available climate data. Those challenges led her to specialize in the field of climate data rescue and quality improvement of historical climate observations. As she has pointed out to young people—especially women—entering the field of climatology, persistence and persuasion have been the keys to her success.

Following Dr. Brunet's lead, I've tried to include a wide range of members in our evaluation committees. From the beginning, I have made a diligent effort to invite our world's many talented women meteorologists onto these committees. For example, that first WMO extreme rainfall evaluation back in 2007 included the gifted Russian female meteorologist V. Davydova Belitskaya.

To counter some of the 'old boys' tradition in science, I also wanted to intermix young and upcoming meteorologists with the older, more established, scientists onto these committees. And that idea has worked. I have found that such intermixing has helped promote cooperation and understanding with *both* the younger and the more senior scientists. For example, as when we undertook an examination of new longest distance and longest duration lightning flashes, I invited onto the committee the then-recently graduated lightning expert, Dr. Daile Zhang who is at the University of Maryland, and she served admirably alongside some of the most respected—but much older—lightning experts in the entire world.

Fourth, we needed 'local experts' on these weather evaluation committees. We needed to have access to the raw data—the actual, unfiltered measurements—from the observer or the device in question. The best person would be someone in the country in which the extreme occurred. For our initial investigation of this twenty-four-hour record, we were fortunate to have Dr. Miguel Cortez of the National University in Mexico accept my invitation to serve on that first inaugural evaluation committee. He was able to assemble a wide range of data, including rainfall data from surrounding events and from other tropical cyclones and information directly from the [Mexican] National Meteorological Service (SMN). Sadly, shortly after his service on my WMO committee back in 2007, Dr. Cortez passed away—may he rest in peace—but his outstanding work has endured.

Fifth, I wanted to include specific meteorology experts who studied the *type* of weather we were investigating. For the twenty-four-hour rainfall event in Mexico, that meant including hurricane experts—scientists who could assess the legitimacy of the rainfall observation based on their very comprehensive knowledge of the phenomenon, in this case, hurricanes. In this first evaluation, that meant contacting one of my good friends, Dr. Chris Landsea of the US Hurricane Center. As I discussed in chapter 1, he was a colleague whom I first contacted back in 2006 about the need of a global database of extremes. He accepted the invitation to serve. In later evaluations, I was fortunate to have other legendary hurricane experts, such as Dr. Jose M. Rubiera Torres from the Cuban Instituto de Meteorologia and Dr. Jack Bevin of the US Hurricane Center, accept membership in my evaluation committees.

As you can see from their very start, these committees were beginning to be crowded. But, nevertheless, I had still one more set of scientists to include—members of the WMO itself. In the beginning that meant the inclusion of the

*Figure 3.2* (Left) Dr. Manola Burent, (center) the Isla Mujeres weather station and (right) the late Dr. Miguel Cortez (National University of México).
Source: Photographs courtesy of Buret, Cortez and the Aspen Global Change Institute.

official who had approved the initial project, my friend and noted climatologist, Dr. Tom Peterson. It also meant that I was honored to invite one of the top people in the entire WMO at the time, Dr. Pierre Bessemoulin, to serve. Back in 2007 prior to his retirement, Pierre was also one of the leading scientists in the national weather service (Météo) of France.

Since that time, the number of people on these committees has fluctuated—once up to a mind-boggling total of twenty-two scientists on a single committee!—and I'll discuss that evaluation a bit later in chapter 13. One thing that I realized was that chairing such an illustrious big group of scientists can, on occasion, be very similar to herding a large and rather uncooperative cluster of cats. Not always an easy task.

So what do these committees actually do? I'll go into a bit more detail about our discussions in the next chapter but, briefly, each committee member receives a set of documents. Those papers describe the equipment, recording procedures and maintenance of the recording station. In addition, in this first evaluation, the raw rainfall totals were available to the committee members.

First, one of the important considerations of WMO verification procedures of weather and climate extremes is—if possible—an initial validation by one or more regional authorities by publication of the record. In essence, did someone in authority publish that this record has occurred? In this particular instance, two regional government agencies did publish this record. The National Meteorological Service, SMN, of Mexico and the US National Hurricane Center had both acknowledged the rainfall record.

When such prior documentation exists, the WMO committee role becomes one of ascertaining whether there are any irregularities in the record that suggest it should not be validated. In other words, the committee's primary role was to determine whether there were any identifiable problems and/or errors with the equipment, procedures or siting locations associated with the record under investigation. In the case of Isla Mujeres, my evaluation committee discussed three potential problems to the record.

First, did anything unusual occur in the manner in which the recording instrument—in this case, a tipping bucket rain gauge—operated during the hurricane's passage? In other words, could the wind or other weather effects have caused a malfunction of the rain gauge? We consulted with experts who work with such instruments. Their opinion was that errors in tipping bucket rain gauge measurements actually tend to be underestimates rather than overestimates. If anything, they said that it is likely the recording amount might have been somewhat less, not more, than the total rainfall that actually fell.

Second, and critically, could there have been other effects that contributed to the high rain amount, such as sea waves splashing into the gauge or water running down wires that could have led to such a high rainfall rate? For this committee concern, we evaluated photographs of the station as well as the rainfall data from surrounding events and from other tropical cyclones. All available records suggested the rainfall amount was credible based on radar data, rainfall observations from surrounding stations and the expected rainfall that should occur with a big, slow-moving hurricane. The Director of the Mexican National Meteorological Service (SMN) identified other independent stations not far away from Isla Mujeres which showed similar high rainfall rates and the Cancun radar animation confirmed that a high reflectivity spiral rain band—likely producing heavy rain—was present over Isla Mujeres for a long time.

In addition, when members of the committee plotted the half-hourly rainfall totals from the station, we saw no abnormal 'spikes' or discontinuities in the twenty-four-hour rainfall totals over the entire day, thereby suggesting no abrupt or marked malfunction of the equipment.

The third potential problem identified by the committee was 'what is the correct twenty-four-hour measurement period for the record?' A previously published rainfall value from this storm (1637.54 mm) was computed from 9 AM local of 21 October to 9:30 AM local of 22 October (actually 49 one-half hour intervals were used rather than 48) and so was totaled for twenty-four-and-a-half-hour period, not twenty-four hours.

Some committee members suggested that we use the 9AM to 9 AM (or using the world-accepted 'Coordinated Universal Time' (UTC)—in essence Greenwich Mean Time—from 00 UTC to 00 UTC value as the true daily twenty-four-hour record. However, since the record is established as a twenty-four hour rainfall rate, not as a 'daily rainfall rate,' we needed to determine exactly which twenty-four-hour period was the record rainfall totaled. When members of the committee analyzed from the raw data, the precipitation value between 9:30 AM Local (12:30 UTC) 21 October to 9:30 AM Local (12:30 UTC) 22 October showed the highest twenty-four-hour rainfall total, an incredible 1,633.98 mm (64.33 inches).

Based on that I, as the WMO Rapporteur of Weather and Climate Extremes, accepted the committee's recommendation. I judged that the highest record (a value of 1,633.98 mm from 12:30 UTC to 12:30 UTC) be used as the official rainfall record for the Western Hemisphere. Consequently, in the World Meteorological Organization's Archive of Extreme Weather and Climate Events, the new record for the twenty-four-hour rainfall in the Western Hemisphere is the value of 1,633.98 mm (64.33 inches) that occurred for Isla Mujeres off the eastern coast of Mexico during the passage of Hurricane Wilma.

With procedures for selecting the 'best of the best' in place, we could undertake a prominent investigation that tested the courage and determination of my meteorologists. What is the highest temperature ever officially recorded for the Earth?

# Interlude: Weather Spies

When I first arrived at Arizona State University back in the early 1980s, I must admit that I experienced a bit of hero-worship. One of the faculty at that time was an extraordinary alpine geographer by the name of Melvin Marcus. A virtual mountain of man—standing well over six feet tall—and a certifiable legend in his field. During his lifetime, Dr. Marcus was an accomplished mountaineer and, I am told, he had logged *fifty* mountain first-ascents. He was—and still is—considered an icon in the area of alpine and arctic geography. As one simple example of his fame, Dr. Marcus had even met such luminaries as Heinrich Harrer. Harrer was the Austrian mountain climber and author who became friends with the Dalai Lama at the time of China's takeover of Tibet (*Seven Years in Tibet*). Yes, Mel loved adventure!

It was one of the thrills of my research life to come to Arizona State University in Tempe Arizona and have a chance to work with Dr. Mel Marcus before he passed away. His death occurred, sadly, back in the late 1990s as he was undertaking alpine field research with students in the snows of Colorado. But—as many of us realized—it was the kind of passing that was what the great Mel Marcus would have wanted, conducting mountain climate research out in the snow fields of the Rockies as he taught a new generation of scientists.

One topic that fascinated both Mel and me was that of the 'Great Game.' The Great Game was a term first used by the great author Rudyard Kipling and is the name that historians use for the clandestine, sometimes violent, and deadly game of political intrigue by the world's superpowers in the nineteenth century.

Much of the Great Game took place in the remote mountains and deserts of Central Asia. For many years, the object of the competition was obtaining critical climatic and geographic information. Daring explorers and spies—the predecessors of James Bond and other modern-day secret agents—were ordered to ferret out vital *environmental* data from inhospitable, unmapped mountain and desert regions.

Russia and Great Britain were the leading players in this pursuit of climatic information. In the early nineteenth century, the British in India grew increasingly concerned about the expansionist plans of the Russian tsars. The de facto rulers of India—the directors of the East India Company—realized that while the renowned British navy protected ocean access to India's riches, an alarming potential existed for land invasion from the north. Yet little was known about the Asian interior. Strict Muslim control and an inbred distrust and fear of *feringhee* (infidel foreigners) meant that Europeans were as likely to be beheaded as welcomed if discovered in these lands.

Despite the grave risks associated with travel in these remote areas, knowledge of the climate and the terrain of Asia's remote interior became a matter of national security. Ambitious military officers, both Russian and British, began secret missions into the Himalayas and Afghanistan, covertly recording information on the passes, the peoples—and the weather.

By the 1860s the Russians had begun a calculated conquest of the Muslim khanates of Khiva, Bokhara and Khokand (today the countries of Uzbekistan, Turkmenistan and Tajikistan). The British in India grew alarmed at Russia's southern encroachment. Could the Russians be considering a move through Tibet into India? And, if so, what routes would their armies travel? The British had no answers to these critical questions for little was known of the eastern Himalayas. Were there traversable passes? What were the distances? Would the weather be tolerable?

More detailed information on this huge unknown area was needed. Yet foreigners were seldom welcomed and often in grave danger. A British officer, Captain Thomas George Montgomerie, had a brilliant solution. He wrote that:

> . . . when I was in Ladakh I noticed the natives of India passed freely backwards and forwards between Ladakh and Yarkant in Chinese Turkestan, and it consequently occurred to me that it might be possible to make the exploration by that means. If a sharp enough man could be found, he would have no difficulty in carrying a few instruments amongst his merchandise, and with their aid good service might be rendered to geography. (Cerveny and Marcus, 1994, 18)

So Montgomerie created an ingenious spy ring of native explorers, each trained in clandestine surveying techniques and known for his intelligence and adaptability. These spies, whose real names were not revealed until they died, were nicknamed 'Pundits,' from the Sanskrit term for a learned man. Before beginning their missions, the Pundits were put through what today would be called a spy school. There, they were drilled on such exotic techniques as pacing at an exact stride (in one case, precisely thirty-three inches) and using Tibetan prayer beads to keep a daily tally of their steps.

They also were given equipment worthy of the fictional Q of the James Bond stories. Their Buddhist prayer wheels held a secret compartment that concealed scrolls of paper for marking routes or noting conditions. A sextant was hidden in the false bottom of the Pundit's traveling chest. The mercury necessary for setting an artificial horizon when taking sextant readings was sealed in one of the multicolored shells used as money in some of the more remote parts of Asia, then poured into a pilgrim's bowl when needed.

A thermometer secreted in a hollowed-out walking stave had two uses: first, of course, to help measure conditions along the trek and, second, to aid in measuring altitude. But how can one measure altitude with a thermometer? Because barometric pressure decreases with height, the nominal boiling point of water decreases 1.8°F (1°C) for each one thousand-foot (305 meter) increase of altitude. The Pundits measured the boiling point of water at different elevations. Indeed, the official inventory of spy equipment given to each Pundit included 'one copper jug and oil lamp for boiling the thermometers' (Cerveny and Marcus, 1994, 18).

The Pundits' measurements have proven surprisingly accurate. They calculated the altitude of the Tibetan capital of Lhasa at 11,700 feet (3566 meters) above sea level. Today, the accepted elevation is twelve thousand feet (3658 meters) above sea level. In evaluating the Pundit's estimate of the capital's elevation, the spy master Montgomerie himself judged that it was only 'some two or three hundred feet in default' (Cerveny and Marcus, 1994, 18).

The Pundit spies did not neglect meteorology in the unexplored region. Pundits found severe weather in Yarkant during the winter of 1864–65. Montgomerie reported that:

> . . . the thermometer early in January [fell] nearly to zero, or 32 degrees below the freezing point. At times the weather was cloudy, and from the 19th to the 26th of January snow fell; but, judging from the general regularity of the observations, on the whole must have been very clear. (Cerveny and Marcus, 1994, 18)

The diligence of the Pundits in recording the weather conditions at Lhasa and other locations was remarkable. One Pundit recorded an amazing 259 separate hourly observations from February to March 1866.

The Pundits' success is undisputed. Yet one of the most bizarre aspects of the whole affair is how the operation's integrity was compromised by spy master Montgomerie himself. In 1866, while the spies were still working, Montgomerie wrote a comprehensive report on the undertaking (discussing aspects of the Pundit's disguise as well as their secret equipment) for the *Journal of the Royal Geographical Society*. In this article he even hinted at the real purpose of the Pundits' travels: 'The progress of Russia in the Ilek Valley seems to be correctly noted; but whether he [the Pundit] is right that the Russians have a fort near Lake Lop ... is very doubtful' (Cerveny and Marcus 1994, 20).

Again in 1868, Montgomerie submitted a second report to the *Journal* in which he gave complete details on the Pundits' illicit spy equipment on a mission to Lhasa. For example, he wrote that:

> Reading the sextant at night without exciting remark was by no means easy. At first a common bull's-eye lantern answered capitally, but it was seen and admired by...curious officials . . . and the Pundit . . . was forced to part with it in order to avoid suspicion. [Consequently] the Pundit was

> at some of the smaller places obliged to take his night observation and then put his instrument carefully by, and not read it till the next morning. (Cerveny and Marcus, 1994, 21)

Discussion of the region's weather also appeared in the second report:

> During the whole time the Pundit was in the Lhasa territory, from September to the end of June, it never rained, and snow only fell once whilst he was on the march, and twice whilst in Lhasa. The snow-fall at Shigatze was said to be never more than 12 inches; but the cold in the open air must have been intense as the water of running streams freezes if the current is not very strong. A good deal of rain falls during July and August about Shigatze and there is said to be a little lightning and thunder, but the Pundit does not recollect seeing the one or hearing the other whilst he was in the Lhasa territory. (Cerveny and Marcus, 1994, 21)

This massive breach of security—and obvious endangerment of the Pundits who were still operating in Tibet—is hard to understand. Montgomerie did make a small disclaimer: 'The two Pundits, being still employed on explorations, their names are, for obvious reasons, omitted." In Montgomerie's defense, it should be noted that the *Journal of the Royal Geographical Society* was not for public sale; it was only distributed to members of the society. The society nevertheless was international—and it included Russian members. It is certain that the tsarist agents in Central Asia found Montgomerie's detailed reports to be interesting reading.

Today that report still provides insight into one of the most bizarre 'weather spy' operations of all time.

# 4 The World's Highest Recorded Temperature, Part 1

> I wonder that any human being should reside in a cold country who could find room in a warm one.
>
> Thomas Jefferson

Heat.

When most of us think of the one weather variable most commonly measured and discussed across the world, the specific quantity that we generally identify first is *temperature* . . . that is, measuring the atmospheric heat of a given location.

Throughout this book, I will be discussing the important instrumentation that meteorologists use to document the weather. Those diverse tools—ranging from thermometers to geostationary satellites—are the backbone of modern meteorology. Therefore, in each chapter I'll delve into the instruments associated with that chapter's particular weather extreme. Let's start with temperature.

Recording devices designed for the measurement of heat have existed since the early 1600s. Many historians credit Galileo Galilei as the inventor of the first instrument to measure air temperature, when he created a device called a 'thermoscope,' although others give credit to a physician friend of Galileo's, Santorio Santorio. Regardless of its inventor, over the next century, the temperature-recording device underwent considerable refinement.

By 1714, the Dutch scientist Daniel Gabriel Fahrenheit invented one of the first reliable thermometers based on the heat expansion and contraction of liquid mercury, instead of using (as was common for that time) alcohol and water mixtures. He chose mercury as his main temperature determinant because it expands/contracts under changes in heat more effectively than other substances such as alcohol or water. For his device, he also developed a temperature scale, which still (somewhat adjusted) bears his name.

For his first device, Fahrenheit selected a scale based on an adult male's body temperature as his highest fixed point (which he identified as 96°F or, in Celsius, a value of 35.6°C). He defined his lowest temperature, based on the heat associated with a mixture of salt and ice, as 0°F (−17.8°C). However, as it was later discovered that body temperature varies, the Fahrenheit scale was later modified to use the boiling point of water as its upper value, 212°F (100°C).

DOI: 10.4324/9781003367956-4

In 1742, Anders Celsius (1701–1744) proposed a scale with zero at the boiling point and one hundred degrees at the freezing point of water (this meant that negatives were not needed except for very high temperatures). A few years later, Linnaeus reversed the scale, creating the first version of the temperature scale that most of the world uses today.

This brief history plays a role in one of the WMO's most celebrated investigations, the evaluation of the world's highest recorded temperature.

Our story begins in 1922 in the land of Libya, at that time under the control of Italy. Libya is a large desert country located in Northern Africa and bordered on the east by Egypt, on the west by Algeria, on the south by Chad and Niger, and, on the north, by the Mediterranean Sea. Following a 1912 war between Italy and the Ottoman Empire in which Italy won, the territory of Libya became known as Italian North Africa.

As part of that control, a series of military forts was established by the Italians across Libya. One of those forts was located at the small town of El Azizia, or sometimes translated as 'Aziziya, located about forty-one kilometers (twenty-five miles) southwest from Tripoli. At that fort on a particularly hot day in September of 1922, a temperature of 58°C was recorded—and, yes, that temperature was recorded in degrees Celsius. Applying the mathematical conversion from Celsius to Fahrenheit, that blistering heat would correspond to a temperature of 136.4°F.

For nearly the next ninety years, that extreme temperature of 58°C would be regarded as the hottest temperature ever recorded on the planet.

Notice that I said 'recorded.' Many social media critics decry modern weather observations particularly when linked to climate change, declaring that the Earth has been much hotter (and colder) than present-day in days prior to instrumentation. That is true . . . but, as I noted earlier, we didn't have recorded measurements of temperature *before* 1600.

But we do know the temperature of the planet prior to the invention of the thermometer!

Temperatures for earlier times in Earth's history have been *estimated* from what are called 'climate proxies'—which are various substitutes for real climate records. Proxies include determining past temperatures using variations in tree rings or ice cores. Such proxies are important and useful measures of long-term climate. However, since proxies are *not* exact measures of climate but rather close approximations, at present they are not used by the World Meteorological Organization as *official* records of past climate extremes.

According to these social media critics, this problem of having actual temperature records only since the 1600s somehow invalidates our thoughts on climate change. 'The Earth has been much hotter (and colder) than present day, so why worry?' they often argue. Of course, what they are refusing to address is the *rate* of change. We are seeing changes in climate occurring within this last hundred years that we haven't seen in the actual (or even the climate proxy) records. I'll talk about that in detail a bit later.

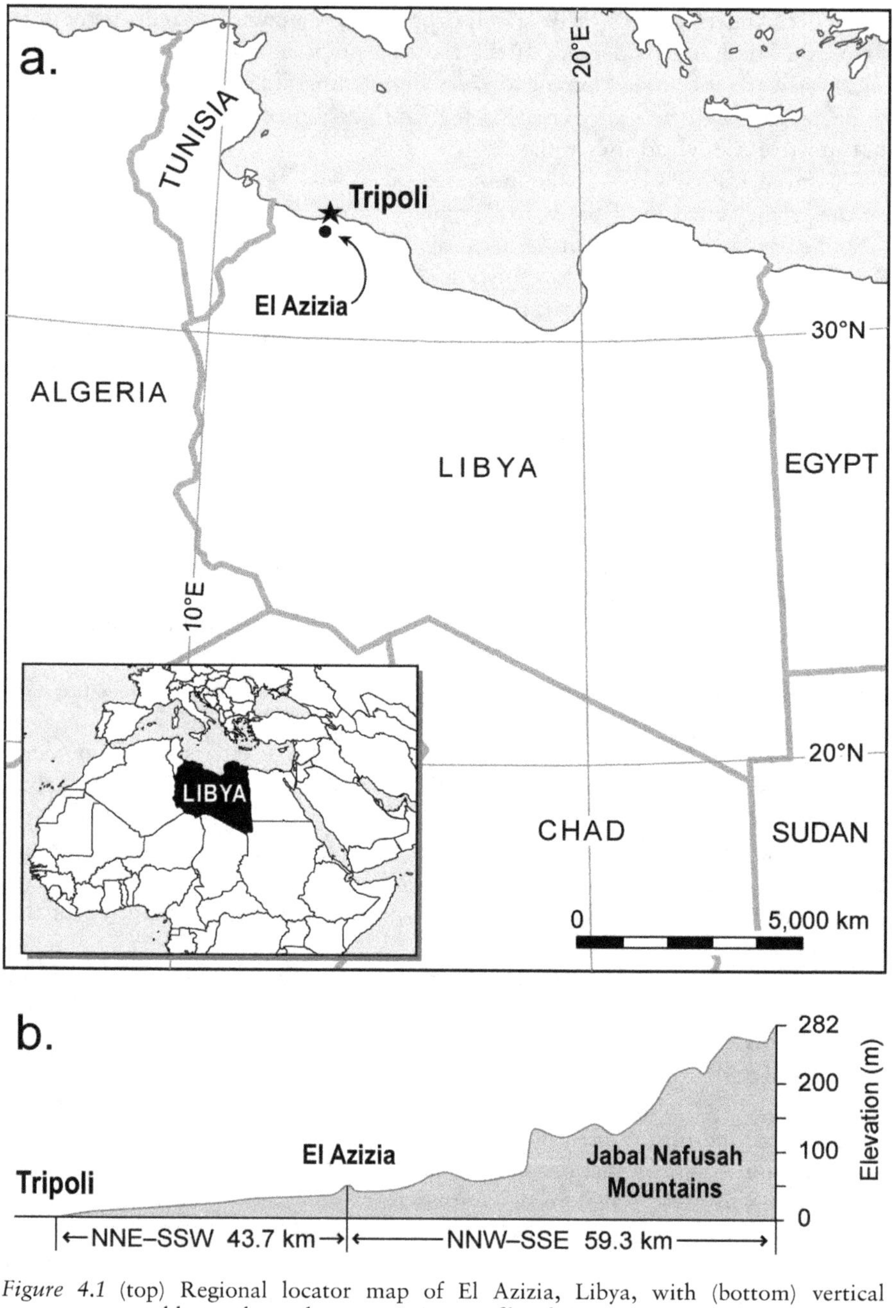


*Figure 4.1* (top) Regional locator map of El Azizia, Libya, with (bottom) vertical roughly north-south cross-section profile of site.

Source: Cartographer, Barbara Trapido-Lurie, courtesy of Bulletin of the American Meteorological Society.

Returning to El Azizia's 58°C temperature, since 1922 that record had long been acknowledged as the highest recorded temperature on the planet. Over the past century, even such illustrious meteorologists of the past century such as C. F. Talman, C. F. Brooks and H. H. Lamb discussed that record. While they and others raised some concerns, the value had been accepted around the planet as legitimate.

Then, one sunny day back in 2010, I received a message from Chris Burt, a good friend of mine. Chris is an accomplished private weather historian who at that time had been associated with Dr. Jeff Master's popular weather company 'Weather Underground' and with NBC television. Chris loves weather records and digging into old weather data. As part of his work (and passion), he has long written an interesting and informative blog about weather and weather records.

Back in 2010, he told me that he had been in correspondence with a climatologist from the country of Libya and Chris strongly believed that El Azizia's 58°C 1912 observation was invalid. Would the WMO be interested in opening an investigation about that temperature observation's validity?

It was an interesting—and important—question.

As I mentioned in the last chapter, we don't open an investigation just for the sake of opening an investigation. Normally, our evaluations begin with a message of a potential new extreme from one of the WMO member nation's weather services.

My evaluation scientists (and myself) work for the WMO as volunteers (and that is true for other work for the WMO, even the well-known IPCC reports). By the way, that volunteerism negates one of the common arguments of climate critics: that all climate scientists have been bought off by special interests. The answer is simply no, we are not bought off. My scientists—and myself—work for free as volunteers for the WMO because we believe that this work is important!

Since my committee members do the work *pro bono publico*, I am hesitant to open a formal WMO investigation without clear, new evidence (a 'smoking gun,' one might say) because I do not wish to waste these important scientists' valuable time. As I have told many people who have submitted requests for investigations, the WMO is *always* willing to examine an old weather or climate record, but such reexaminations are best performed *if* important new evidence about an observation's legitimacy has been uncovered.

In this case of El Azizia's 58°C 1922 observation, Chris informed me that we had such new evidence in spades. He told me his colleague in Libya had uncovered the *actual* handwritten weather observation sheet from 1912! That genuine nearly-hundred-year-old weather log sheet written by the individual in charge of the weather station at El Azizia showed an unmistakable error in the 58°C observation.

Why was this important? The problem with evaluating old weather records is often incomplete information. For old records, we can't call up the observer and say, 'What happened?' We are forced to become weather detectives to assess the available (often incomplete) evidence. Like Sherlock Holmes, we try to determine what really happened using the available existing information. Having the original weather observation sheet from 1922 was the weather equivalent of having a photograph of the murder scene.

Based on that (and other information which I discuss below), I thought that there was ample cause for opening an investigation of the El Azizia observation. When others at the WMO agreed, we began the interesting process of selecting a committee.

First, as I discussed in the previous chapter, for this investigation we needed a knowledgeable, international, and recognized committee.

Who should I invite? The question wasn't mine alone—by this time in 2010, several others in the WMO were offering opinions and making considered suggestions to me. I needed to take those ideas into account as I made my selections.

For our first member, I requested the climate historian Chris Burt be a member of the evaluation committee. Yes, Chris came into the investigation with the preconceived firm stance that El Azizia's 58°C temperature was invalid. Yes, that might be problematic. As scientists, we should always strive to remain unbiased and judgement-free until the evidence convinces us. But in this case, I felt (and argued) that Chris's initial discovery work had been important. I wanted him on the committee and, frankly, he deserved to be on the committee. After some discussion, I convinced the necessary WMO leadership on that point.

Representatives from many other counties also needed to be invited.

Back in 1922, Libya had been under the control of the country of Italy. The actual observation of the 58°C temperature had been made by Italian soldiers. Twenty-five miles south of Tripoli, the weather station at El Azizia was established in April 1913 by an Italian physical scientist (Dr. F. Eredia) but later the site was used as a military station until 1926.

So, getting specific representation on the committee from the country of Italy was important. Their national weather service operates within the framework of the Italian air force. Through diligent cooperation between the WMO and the Italian air force, we received the services of two distinguished climatologists from that country, Lieutenant Colonel Gianpaolo Mordacchini and Colonel Vinicio Pelino.

I was also pleased to invite my good friend, Dr. Manola Brunet, who was (and still is) the Director for Centre for Climate Change, Tarragona, Spain (and is now one of the top people at the WMO), and Dr. Fatima Driouech, who at that time was at Morocco's Direction de la Météorologie nationale (their national weather service). I also invited the illustrious Dr. Pierre Bessemoulin of the French national weather service Météo; Jose Luis Stella, a talented meteorologist from Argentina; Dr. Philip Eden, a British weather historian of the Royal Meteorological Society (who, quite sadly, has passed away since that time); Dr. David Parker of the British Met Service and an expert experienced with Middle East weather; and Dr. M. M Abdel Wahab of the Department of Astronomy and Meteorology at the Cairo University in Egypt.

And, of course, Chris's most valuable contact from Libya needed to be invited onto the committee. That was my first introduction to (now, Dr.) Khalid Ibrahim El Fadli.

At the time Khalid El Fadli was the Director of Climate & Agrometeorology Directorate at the Libyan National Meteorological Center (LNMC). A key factor with that appointment was that in 2010 the LNMC was a part of the government run by Colonel Muammar al-Gaddafi. Was that a problem?

Believe it or not in this time of hyperpolitics, weather doesn't have political affiliations—it just happens. Having Khalid on my extremes evaluation committee with his Gaddafi affiliation was *not* a concern for the United Nations or the WMO. However, before the end of the investigation, because of that affiliation with Gaddafi, Khalid's participation in our evaluation would be tested to limits that most people—and meteorologists—would have never imagined.

All those invitees accepted their invitations. Still today, I shake my head in amazement as I recall those people. This was an illustrious, international team of men and women. This group of talented meteorologists and climatologists made the investigation of that 1922 world-record 58°C temperature much easier than it might have been.

We began our work in late 2010.

Before we could proceed too deeply into our investigation, world events forced our evaluation to take a quite unexpected turn.

Beginning in the early months of 2011, a deadly wave of political unrest swept across the entire Middle East. Historians would later call that turmoil 'the Arab Spring' and it would lead to the overthrow of several governments throughout the region.

For our WMO investigation, my committee of scientists and I were watching the growing unrest in one country closely.

Libya.

Early in 2011, major protests broke out against the Libyan government run by Colonel Gaddafi. The embattled leader proclaimed that any rebels working against the government should be 'hunted down street by street, house by house and wardrobe by wardrobe' (Kawczynski 2011, 242–43). A short time after this proclamation, Gaddafi's army opened fire on protesters in the eastern city of Benghazi, killing hundreds. The slaughter spurred massive outrage, both nationally and globally. Fueled in part by the massacre, the rebel uprising spread throughout the eastern half of Libya. By the end of February 2011, the rebels, who now called themselves the National Transitional Council (NTC), had seized control of the eastern cities of Benghazi, Misrata, al-Bayda, and Tobruk.

In March 2011, the United Nations Security Council declared a no-fly zone over Libya. The aim was to protect the civilian population from aerial bombardment. The UN called on its member nations to enforce the no-fly zone as NATO began aerial enforcement of the UN resolution. On 30 April, a NATO airstrike killed one of Gaddafi's sons and three of his grandsons in Tripoli.

By July 2011, over thirty governments around the world had recognized the rebel NTC as Libya's legitimate government. With such backing—and the

*Figure 4.2* The WMO committee members of the 1922 Libyan Temperature Extreme (top to bottom, right to left): Dr. Khalid El Fadli, Randy Cerveny (the author), Christopter C. Burt, Philip Eden (deceased), David Parker, Manola Burnet, Thomas C. Peterson, Gianpaolo Mordacchini, Yinicio Pelino, Pierre Bessemoulin, José Luis Stella, M. M. Abdel Wahab and Fatima Driesouch.

Source: Courtesy of the individuals and the WMO.

enforcement of the UN's no-fly zone—the rebel militia continued to push westward and, by late summer, attacked Libya's capital of Tripoli.

During these events, my WMO evaluation committee and I were quite concerned as we listened to the distressing news from that war-torn country. Remember that our committee's local contact, Khalid El Fadli, was working at the Libyan National Meteorological Center in the capital of Tripoli—then under the control of the strongman Muammar Gaddafi.

For a very long time, we heard nothing from Khalid El Fadli.

Was he safe? Was he even alive?

Then—finally—a quick short message. Dr. El Fadli and his family were safe, having fled to the other side!

In September, he wrote to us:

> . . . our situation, during the Battle of Tripoli, our work premise has been stolen and destroyed and we have faced some difficulties to access with internet right now I'm using my home connection, [a] very low one. But in some coming days the situation will be more stable and I'll inform accordingly.

He later wrote:

> During that critical time [in the summer of 2011] all communication systems in Libya were shut down by the [Gaddafi] regime so it was impossible to communicate with anyone, even inside the country. Mobile telephone communications were restricted, and even local calls were controlled and monitored. What was amazing however, believe or not, was that my office satellite internet connection was still up and running. But using such posed serious dangers, if anyone discovered me, I would probably lose my life. Hence, I never used that connection.

Incredible courage! This is still another one of the (surprisingly) *many* times that I have heard of a meteorologist in fear for his actual life for simply doing his job!

Wait a minute? You don't think meteorology can be a dangerous job? Sorry, that isn't true. Studying weather can be *quite* dangerous! For example, consider the brave meteorologists of the famed 'hurricane hunter' flights. Every time they take off to fly into those deadly storms, they are flying straight into danger! I'll talk more about them in a later chapter.

But to return to Dr. El Fadli and his own life-threatening situation, his first-hand account of that year (2011) in Libya is harrowing:

> The first three months (February–May),' Khalid said, 'I was able to reach my office (my home being about five kilometers [three miles] east of El Azizia and forty km [twenty-five miles] to my office in Tripoli) but then in May we suffered from short fuel supplies, electricity, and even cooking gas. You can imagine how difficult our lives became!

As the summer progressed and the rebels began their offensive against the Gaddafi government in Tripoli, Khalid's life became even more perilous.

He wrote that once, near the beginning of the climactic Battle of Tripoli:

> . . . when I borrowed a car belonging to the local United Nations office (since I had no fuel for my own car) I was driving to morning prayers (04:00 am) with my sons and suddenly we came under gunfire from the back and rear of the vehicle. The vehicle was struck as I drove at a crazy speed with our heads ducked low. Our life was spared by the grace of God. This happened in late July.

Up to that point in my career, I must admit that I had never thought one of the perils of meteorology would actually be getting *shot at*! Fortunately, Khalid and his family survived that deadly encounter.

Finally, in September of 2011, our heroic Libyan meteorologist wrote to me that:

> . . . gradually the situation will be improved and next week [I] will come back to my office and try to re-manage how to communicate with you and resume our activity in general so that after next week the situation will get normal.

Normal?!

I still shake my head in wonder at Dr. El Fadli's words. This brave statement was coming from a guy who—quite literally—had been shot at a few weeks earlier for simply doing his job. Incredible courage!

Dr. El Fadli proved true to his word. Soon after that message reached me in September of 2011, Khalid returned to his job at the Libyan National Meteorological Center, but this time now working for the new rebel government. And with his return, he had regained his access to all the old weather records, including those from 1922.

We were back in business!

The evaluation of the 1922 world-record 58°C temperature could resume!

# Interlude: Freaks of the Heat, Part 1

We live in times of increasing heat. Sorry, if that offends you. That is a fact, not an opinion.

Think of the number of times in recent years that your own particular area on Earth has experienced a record high temperature, as opposed to a record low temperature. For example, in my location—Phoenix, Arizona—the last time we recorded a daily record low temperature was more than twenty years ago—as opposed to new daily record highs being recorded almost every year.

In simple words, we are getting hotter. Another way to view that disturbing fact is that heat now kills more people across the planet than *any other* weather-related phenomenon—including tornadoes and hurricanes combined. Indeed, some of my valued colleagues at Arizona State University have invested their entire careers into studying the many effects of increasing heat on our urban landscape—and, particularly, on people. I'll talk more about that in the last chapter.

But studying the consequences on high temperatures on people raised an interesting question.

How long have we been scrutinizing the heat's effects on people?

The answer, surprisingly, is a very long time.

In 1775—the time of the American Revolution—a British researcher named Dr. George Fordyce conducted a bizarre set of experiments to 'scientifically' test the limits of human endurance to increasing heat. Fordyce invited a fellow scientist, Charles Blagden, and a few other colleagues to observe the strange experimentations.

According to Blagden, Fordyce first:

> . . . procured a suite of rooms, of which the hottest was heated by flues in the floor, and by pouring upon it boiling water; and the second was heated by the same flues, which passed through its floor to the third. The first room was nearly circular, about ten or twelve feet in diameter and height, and covered with a dome, in the top of which was a small window. The second and third rooms were square, and both furnished with a sky-light. (Blagden 1775, 112)

Blagden's account of Fordyce's first experiment (with all temperatures in Fahrenheit):

> In the first room the highest thermometer stood at 120°, the lowest at 110°; in the second room the heat was from 90° to 85°; the third room felt moderately warm, while the external air was below the freezing point. About three hours after breakfast, Dr. Fordyce having taken off all his cloaths [clothes], except his shirt, in the third room, and being furnished with wooden shoes, or rather sandals tied on with lift, entered into the second room, and staid [stayed] five minutes in a heat of 90°, when he began to sweat gently. He then entered the first room, and stood in the part heated to 110°; in about half a minute his shirt became so wet that he was obliged to throw it aside, and then the water poured down in the streams over his body. Having remained ten minutes in this heat of 110°, he removed to the part of the room heated to 120°; and after staying there twenty minutes, he found that the thermometer placed under his tongue, and held in his hand, stood just at 100°, and that his urine was of the same temperature. His pulse had gradually risen till it made 145 pulsations in a minute. The external circulation was greatly increased; the veins had become very large, and an universal redness had diffused itself over the body, attended with a strong feeling of heat... (Blagden 1775, 113)

In future experiments, Fordyce really cranked up his makeshift sauna's heat. As Blagden recorded:

> The honourable Captain Phipps, Mr. Banks, Dr. Solander, and myself, attended Dr. Fordyce to the heated chamber, which had served for many of his experiments with dry air. We went in without taking off any of our cloaths. . . . When we first entered the room, about 2 o'clock in the afternoon, the quicksilver [mercury] in a thermometer which had been suspended there stood above the 150th degree. By placing several thermometers in different parts of the room we afterwards found, that the heat was a little greater in some places than in others; but that the whole difference never exceeded 20°. We continued in the room above 20 minutes. . .
>
> Upon entering the room a third time, between five and six o'clock after dinner, we observed the quicksilver in our only remaining thermometer at 198°: this great heat had so warped the ivory frames of our other thermometers that every one of them was broken. We now staid [stayed] in the room, all together, about ten minutes; but finding that the thermometer sunk very fast, it was agreed, that for the future only one person should go in at a time, and orders were given to raise the fire as much as possible. Soon afterwards Dr. Solander entered the room alone, and saw the thermometer at 210°; but during three minutes that he staid [stayed] there, it sunk to 196°. Another time, he found it almost five minutes before the heat was lessened from 210° to 196°. Mr. Banks closed the whole, by going in when the thermometer stood above 211°; he remained seven minutes, in which time the quicksilver had sunk to 198°. . . These experiments, therefore, prove in the clearest manner, that the body has a power of destroying heat.

And Fordyce's *ultimate* experiment involved cooking a steak in the oven-like heated chamber ... *with a man present.*

The heat scientist himself stayed:

> . . . for a considerable time, and without great inconvenience in a room heated by stoves to 260°F. The lock of the door, his watch and keys lying on the table, could not be touched without burning his hand. An egg became hard; and though his pulse beat 139 times per minute, yet a thermometer held in his mouth was only 2° or 3° hotter than ordinary. He perspired most profusely, and it was the cold generated by the evaporation of this moisture which enabled him for a short time to bear a temperature that would have cooked a beefsteak. (Blagden 1775, 121)

Another scientist, Dr. Douglas H.K. Lee, further described that last curious set of experiments by Fordyce:

> Men remained in this atmosphere [for] fifteen minutes, without any noteworthy rise in body temperature, while a beefsteak was nicely cooked in thirteen minutes. These observers noted that water kept in a bucket did not boil, even though left in for some time. They failed, however, to draw the conclusion that man's failure to follow the beefsteak was of a kind with the failure of the water to boil; that evaporation provided the means of heat regulation. (Lee 1948, folder 4)

Yes, it can be humbling to realize that scientists are just people . . . and so they can occasionally be quite weird in their experiments.

Before I leave this interlude on 'freaks of heat,' I must also note that heat perception can also be psychological in nature. As proof, I offer this somewhat droll anecdote from 1963, involving then Vice President Lyndon Johnson. In the early years of the Kennedy administration and to his flight crew's dismay, the vice president would often become grouchy regarding the plane's cabin temperature. Johnson had developed the persistent habit of berating the flight crew about lowering or raising the temperature onboard the vice president's official plane, Air Force Two.

After a while, the crew became tired of his scolding, so they engineered a rather ingenious solution. They installed a fake temperature control—not linked to anything else—in the plane's Conference Room, which they said would allow the vice president to directly control the temperature of the plane. Johnson was so pleased with his supposed ability to manipulate the controls himself that he didn't even notice when the temperature remained exactly the same.

For the harassed crew, their ingenious psychological fix worked. The vice president stopped complaining.

# 5 The World's Highest Recorded Temperature, Part 2

> If I owned Texas and Hell, I would rent out Texas and live in Hell.
>
> General Philip H. Sheridan, on an 1866 visit to the state of Texas

As I detailed in the last chapter, my WMO committee, which in 2010 had started an examination of the old 1922 world-record temperature of 58°C in El Azizia Libya, had ground to a sudden stop. The entire investigation had been shelved throughout the first half of 2011 as a bloody revolution erupted across that African country. Throughout that time, we waited anxiously for word from our on-site atmospheric scientist in Libya, the courageous Khalid El Faldi. Then, in late August of 2011, we heard from him. He had miraculously survived the overthrow of his country's government. Dr. El Faldi had literally fled the fighting with bullets flying around him and his family.

That survival meant that the WMO evaluation committee could get back to its important work!

The question now was: How would this group of learned scholars evaluate the 1922 Libyan 58°C temperature?

First, committee member and climate historian Chris Burt and I had completed a short formal brief for the committee that detailed the basic background of the 1922 Libyan 58°C temperature. It contained numerous maps, the past history and analyses of the event, photos of the type of instrument used and even a comparison of the temperatures at that station compared to surrounding stations.

Based on their analysis of that background report, I asked this evaluation committee—as I do with all my committees—to address five questions regarding any potential new weather extreme (or, as in this case, an old weather extreme):

a Is there need for more raw data or documentation on this event to determine its validity or invalidity? Are there other data or other analyses corresponding to this time/place extreme event?
b Are there any concerns about the equipment, calibration, measurement procedures, or other processes/procedures associated with the measurement of the event?
c Are there any concerns associated with the nature of the event that would raise questions regarding the validity of the record?

DOI: 10.4324/9781003367956-5

d Are there any other concerns associated with the event?
e Fundamentally, does the documentation support or refute this current world weather record?

In this case, alarm bells began ringing for all five questions as we sifted through the evidence.

In our careful and very detailed evaluation of the available 1922 data, my WMO extremes evaluation committee identified not one but a total of *five* major concerns with this extreme record. We ascertained these specific points: (a) a potentially problematical instrument, (b) a probable new and inexperienced observer at the time of observation, (c) an unrepresentative microclimate of the observation site, (d) a poor correspondence of the extreme to other records and (e) a poor comparison to subsequent temperature values recorded at the site.

Let's take each of those concerns.

First, we asked the question: 'Was the right kind of equipment employed?'

In this case, the type of thermometer used for the extreme measurement was a Bellani-Six thermometer. It was a replacement instrument to the standard thermometer used for observations at the weather station. Why? A few days before the legendary 58°C observation back in 1922, the army base had come under attack. And in that attack, the base's standard thermometer had been shattered, and the normal weather observer had been incapacitated or killed.

Although we do not have a photograph of the actual instrument used, the Bellani-Six thermometer likely resembled the one depicted in Figure 5.1. In Italy, this instrument, first created by James Six (1731–1793), was called a Bellani-Six, as Angelo Bellani was the first Italian manufacturer of such instruments. This instrument was not designed as a professional monitoring

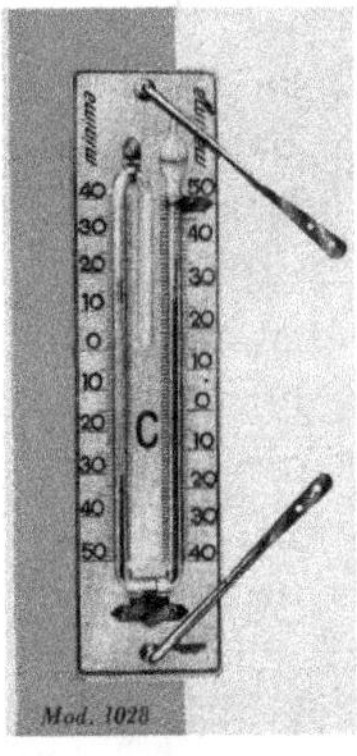


*Figure 5.1.* A 1933 instrument catalog image of the Bellani-Six style thermometer
Source: Image supplied by Paolo Brenni, President of the Scientific Instrument Commission, and courtesy of Library of the Observatorio Astronomico Di Palermo, Gisuseppe S. Vaiana, and the *Bulletin of the American Meteorological Society*

instrument. Several experts informed us that this type of thermometer was used in private households rather than employed as official recording instruments.

As a strange instrument, the Bellani-Six thermometer used *both* alcohol and mercury to make its measurements. It consisted of a U-shaped tube with a sealed alcohol tube of mercury and a sealed ampoule for the alcohol. The ampoules and the linking tubes contained alcohol, while mercury was in the middle of the tube. Steel pins within the alcohol were pushed by the mercury and, when the temperature changed, those pins held their positions. Their lower ends indicated the minimum and maximum temperatures recorded. Daily reset of the instrument was accomplished by use of a magnet.

Judging from notes by the Italians in charge of weather observations at the time, it is unlikely that the Italian Colonial Meteorological Service did not use such an instrument as a professional recorder. It is probable that the Bellani-Six instrument was found elsewhere in the military fort and put into operation when the official maximum thermometer was broken.

In our hunt for information, my committee discovered that a respected instrument meteorologist, W. E. Knowles Middleton, had written that:

> The difficulty with the Six's thermometer, and indeed with all thermometers containing both spirit [alcohol] and mercury, is that the spirit wets the glass and can at length pass between the glass and the mercury [leading to error in readings]. This was clearly recognized by the middle of the 19th century and led to the abandonment of such thermometers as serious meteorological instruments. (Knowles Middleton 1966, 159)

Based on his extensive knowledge of weather instrumentation, a committee member, Philip Eden of the UK's Royal Meteorological Society, confirmed the Bellani-Six thermometer's unreliability. In particular, he noted that a critical part of the instrument casing—a slide that indicated the temperature—covered a length on the thermometer equivalent to 7°C. He suggested that it would be easy to mistakenly read the *top* of the slide for the daily maximum temperature rather than correctly reading its *bottom*, especially if an inexperienced observer recorded the measurement.

Okay, the committee believed that the wrong instrument was used.

Second, the observer's potential inexperience raised concerns for my evaluation committee. My experts suggested the high probability that a new and/or inexperienced observer started recording at the Azizia site, beginning on 11 September 1922, and likely misread that difficult-to-use Bellani-Six thermometer.

This was established by our 'smoking gun,' the original weather datasheet from 1922. The original data entry log shows an abnormality—a change in handwriting and in log-keeping practices—beginning on 11 September 1922 and continuing throughout the month, when the daily temperature maximums and minimums were misplaced in adjoining columns on the log, indicating that the observer was not familiar with the recording process. According to our Italian members of the WMO committee, the observer would have been associated with the Italian military but, unfortunately, no specific name, rank or other identification of that individual exists.

Strike two, a probable inexperienced observer.

Third, the microclimate of the weather site was not typical of the area in several aspects. The temperature observations were made over a black concrete-coated plaza of a small military fort on a hill. The plaza coating of tarred concrete could amplify surface heating beyond the norms for a natural desert environment. After the instrument shelter in El Azizia was relocated in 1927, only two other temperature readings above 50°C (in the ensuing 48 years of record) were measured at the site.

Strike three, a bad location for the weather measurements. In baseball, with three strikes, someone is declared out but let's be really fair. Let's give our 1922 58°C temperature two more opportunities.

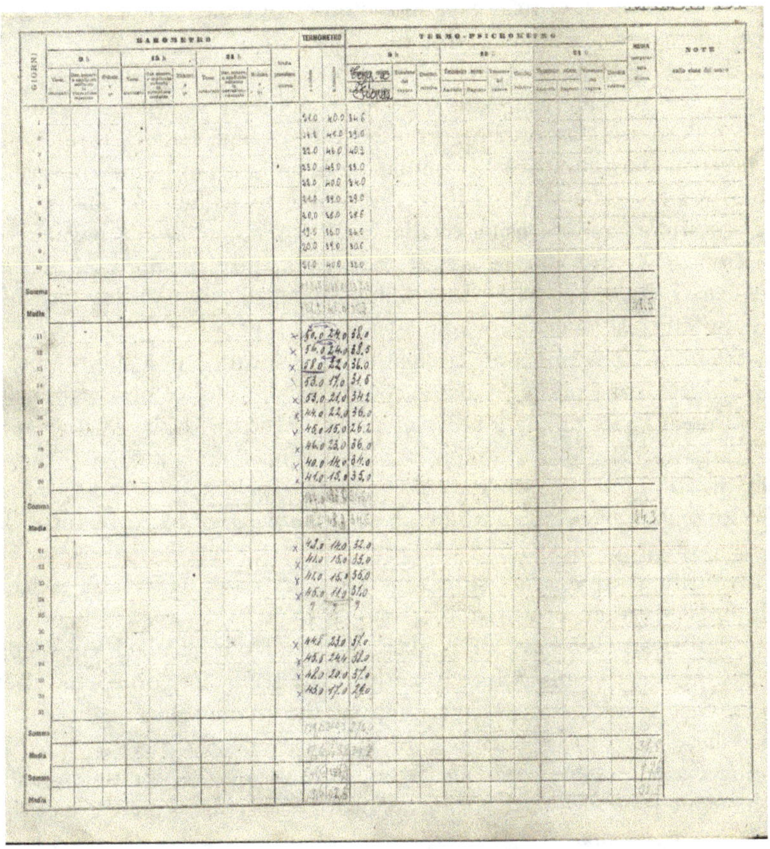


*Figure 5.2* A portion of the original meteorological observation sheet for El Azizia for the month of September 1922. Note the mislabeling occurring in the maximum/minimum temperature columns starting on 11 September 1922 ('x's and ink marking added at an indeterminate time likely long after the original daily observations)

Source: Photo courtesy of the *Bulletin of the American Meteorological Society*

Fourth, was the 58°C value at Azizia consistent with surrounding values? Although, by definition, a record value should be an extreme that is not exceeded by any surrounding values, we would expect that its values should be reflected in similar near-record high temperatures around the area. One way that we have evaluating that is through a relatively new tool used in climate studies called *reanalysis*.

Reanalysis is a clever way to employ modern weather models—the same type of models that make the forecasts that you see every night on the evening news—to examine times when actual weather studies weren't possible. Remember that weather observations have always been unevenly distributed across the globe, and that we haven't always had the abundant weather information that we do today. Of course, even with modern satellites and weather balloons, observations alone cannot provide a complete and accurate picture of the state of the earth system across the globe at a given point in time. Reanalyses allow us to fill in critical gaps in the observational record, and they do so in a manner that is consistent across the globe.

How does a reanalysis weather model work?

Suppose, for example, we want to look at the likely weather patterns for 11 September 1922, the day of the purported Libyan high temperature. What we do is take an accurate weather model and initialize it—start it—using all available weather data that we have been able to gather for that particular instant. Now, of course, using available 1922 data, it won't have as much information as we are able to collect today, but, through the use of complex calculus-based equations and those initial data, the model *interpolates* the probable weather across the entire planet for 11 September 1922. It even extrapolates the most likely upper air circulation—the location of a subtropical high-pressure system, for example—that accounts for the surface observations.

But with reanalysis simulation, we aren't concerned with the weather model's actual forecast, instead we simply want its 'time = 0' analysis—the exact weather situation just *before* we would begin making a normal weather forecast model. Consequently, we aren't looking at a real weather forecast. Instead, it is the starting situation for making a weather forecast. For a reanalysis, we collect all of these time = 0 weather situations for every day as far back as we have enough initial weather information.

Using reanalysis, we have been able to extend our detailed knowledge of actual weather conditions back to as early as the 1850s! Of course, the scientific quality of that 'hindcast' does deteriorate as the amount and quality of initial data is reduced the further back in time we go. Even so, reanalysis provides an interesting extra tool for studying the weather conditions for times when we didn't have modern weather instruments or for areas where there aren't enough actual weather stations.

For our evaluation, one of our esteemed committee members, Dr. David Parker of the UK Met Office, conducted a reanalysis of surface conditions across the Libyan region for September 1922. The results that he obtained were noteworthy. He found that the difference between the temperature from the reanalysis's reconstructed weather patterns versus the purported temperature at Azizia was *six standard deviations* different!

What that means, statistically, is that only 0.0000002 percent of Azizia's temperatures ever should fall outside of the expected results—that is the set of temperatures generated by the mathematical reanalysis. So, if the purported 58° C temperature was real, reanalysis results suggest such an occurrence would be close to miraculous—unimaginably rare. While this doesn't conclusively prove that the 58°C couldn't have happened, it strongly points in that direction.

Additionally, remember that reanalysis can also give us the probable weather patterns above the surface for any time where we have initial data. When we look at the upper air circulation above Libya for 11 September 1922, Dr. Parker found that the lapse rate—the change in temperature with height—would have been physically implausible. Again, this doesn't conclusively discount the reality of a 58°C observation but it does add weight to the possibility of error in that observation.

Strike four. The available mathematical analysis of the day doesn't support a surface temperature of 58°C.

Fifth and finally, beginning on 11 September 1922, the maximum temperature readings increase dramatically while the minimums continue more or less within a fixed range. The daily excursions of temperature skyrocket, with the 24 September 1922 diurnal temperatures ranging from 11°C (52°F) to 45°C (113°F). Although a 34°C (61°F) daily change from low temperature to high temperature is possible, it does indicate a major shift from the norms, which were established before and after the event. Note that this shift occurred at the same time as the new observer likely began taking observations.

Strike five, an abrupt change in the time series of temperatures at the Azizia site occurring right at the time of the observer change.

What do these five concerns mean?

The WMO evaluation committee concluded the most compelling scenario for the 1922 event was that, beginning on 11 September 1922, a new and inexperienced observer took charge of the weather observations at Azizia. Such an individual was not trained in the use of an unsuitable replacement instrument, a Bellani-Six thermometer that can be easily misread. That observer mistakenly recorded the observation using the wrong end of the recording pin and was off in the observation by about seven degrees Celsius. Such a scenario is consistent with all five of the problems identified by the committee.

Given this, can we entirely dismiss the 58°C observation?

Of course not. Because we don't have a time machine, no *absolute* determination of the extreme can be made at this late date. But my WMO panel of experts unanimously determined that, based on the five areas of concern that I detailed above, there was *sufficient* evidence to invalidate the temperature extreme of 58°C at El Azizia as the world's official highest recorded temperature.

The situation is similar to a court of law. We may not have a video of the exact crime, but if the available evidence supports the conclusion that the crime did occur, then the jury can render a guilty verdict. In essence, my committee was following in the footsteps of the great Sherlock Holmes, piecing together the available clues of the weather situation back in 1922 to arrive at the truth. And the truth in this case was that it was likely the 58°C observation was made in error.

Therefore, in my determination as the WMO Rapporteur of Weather and Climate Extremes, I invalidated the El Azizia measurement. That meant the second highest temperature in our archive then became the first. In other words, in the World Archive of Weather and Climate Extremes, the new official highest temperature recorded on the planet is 56.7°C (134°F) and was measured on 10 July 1913 at Greenland Ranch (Death Valley), CA, USA.

But what if that old 134°F Death Valley temperature in 1913 also isn't correct? If there was one bad observation made in Libya in 1922, then perhaps all old historical extreme weather records are bad.

Well, there are some meteorologists who believe that.

For example, my good friend Chris Burt and others have tried to make a case that the observer in Death Valley knowingly entered a bad observation of 134°F into the logbook back in 1913 and that temperature extreme should also be invalidated. He and a colleague have written an interesting blog about the potential inaccuracy of the 1913 observation. Is he right?

Regarding the WMO's current position on this, there are two things that preclude me from opening an investigation on the Death Valley observation. First, we like to have a smoking gun—some new evidence that suggests the observation was wrong. In the Libyan case, that was the original 1922 log sheet that demonstrated the observers changed before the 58°C observation.

In contrast, we do *already* have the original 1913 observation sheet from Death Valley—I've seen it—and there doesn't appear to be any obvious problems in the observation as written on that log sheet. Indeed, one of last century's top American meteorologists from California, Dr. Arnold Court, stated in a published academic article in 1949 that 'there is no indication that the readings as reported were not correct' (Court 1949, 220). Yes, some climatologists have considered Court to be cantankerous, but his work was superlative. I even had the distinct honor of meeting him and having him favorably comment on some of my early work, many years ago.

Second, generally we at the WMO want a country's National Extremes Committee or equivalent—if they have one—to be first judge on an existing extreme within its own borders. In the Death Valley case, that means the US National Climate Extremes Committee (mentioned in the first chapter) would normally be first to judge any new evidence. Currently, that US committee hasn't reevaluated the 1913 Death Valley temperature.

So, currently, two facts are clear:

First, to the best of our scientific ability, the temperature of 58°C reported in September of 1922 in Libya was inaccurate due to a variety of specific concerns. Second, at this time, the official highest temperature for the world is 134°F (56.4°C), measured in Death Valley, USA in July of 1913.

With regard to that second fact, a key phrase is 'at this time.'

Since 2012, when we announced the invalidation of the Azizia 58°C observation, we have had several new temperature evaluations.

In particular in 2019, we finished an investigation of not one but two near-global-record high temperatures of 54°C (129.1°F), one in Mitribah, Kuwait on 21 July 2016 and a second in Turbat, Pakistan on 28 May 2017. That WMO investigation—one of the most extensive evaluations of a temperature observation ever undertaken—involved both temperature sensors being flown to an independent calibration laboratory in Italy (Istituto Nazionale di Ricerca Metrologica).

In that precision testing facility, the two sensors were meticulously checked to ensure the temperature that was reported by the instrument was the true observed temperature. As part of the analysis, the two sensors were placed in a sealed and controlled chamber that could be exactly maintained at specified temperatures.

With incredible precision, we found that the Mitribah temperature was likely 53.87°C ±0.08°C, while the Turbat temperature was most probably 53.72°C ±0.04°C. I have accepted those two temperatures as the third and fourth highest temperatures ever recorded for the planet.

At the time of this writing, I have WMO committees evaluating new temperature extremes at Death Valley in 2020 and 2021 that may rank as the third hottest temperature observation for the globe. Also, in Death Valley within the past couple of years, the National Weather Service has established a new official weather station located at what is called 'Badwater Basin,' a place that is at the lowest possible elevation (282 feet or eighty-six meters below sea level) in all of Death Valley. An important physical fact: All other factors remaining constant, air gets hotter at lower elevations. Therefore, this new station will be recording hotter temperatures than its counterpart at Furnace Creek Ranch about fifteen miles away.

Remember what I said in the last chapter: 'We are seeing changes in climate occurring within this last hundred years that we simply haven't seen in the actual (or the climate proxy) records.' I also mentioned that, in recent years, most areas on earth have experienced many more record high temperatures, as opposed to record low temperatures.

In simplest words, we are getting hotter.

What that means is that we *will*—not might, but *will*—exceed that record of 134°F (56.4°C) in the Death Valley back in 1913.

It is only a matter of time. And probably only a short time. . .

# Interlude: Freaks of the Heat, Part 2: 'Dam Weather'

Since I ended the last chapter talking about Death Valley, USA, let's continue discussing the American Southwest and its incredible heat.

One of the iconic examples of modern engineering is the southwestern marvel known across the world as Hoover Dam. It stands as a working monument to the ingenuity and fortitude of our forefathers to overcome the depths of the Great Depression and harness the forces of nature. But few people realize how much of a role weather has played in both the dam's construction and in its current and future operation. One of the most poignant aspects of 'dam weather' was in the 1930s, during the rapid construction of Hoover Dam. That construction occurred during one of the hottest times in the Southwest's entire weather history.

With the stock market crash in 1929 and the following depression, hungry and desperate people looked to the federal government for relief. For many, that relief came in the opportunity to work on Hoover Dam. By 1931, thousands of despairing families moved to southern Nevada in hopes of finding jobs on the project. That migration led to the creation of makeshift settler camps, including Boulder City, a few miles west of the dam site.

Life in those camps was close to intolerable. The words of some of the inhabitants poignantly describe the situation:

> Bob Parker: [The mess hall] was a very well-laid kitchen, except that it was very hot down there that summer [1931]. The temperature got to 112 [°F, 44°C]. It was the coolest it got at five o'clock in the morning for six weeks straight. In that kitchen, with all those big ranges and cookstoves, it was a fright in there. . .
>
> We lost four down there because of the heat in the canyon. Anderson's lost four men from heat prostration. . . When somebody there in the canyon became overcome with the heat, we dashed out there with these ice buckets and we'd pack them in the ice. If their heart took it and they survived, OK. But if their heart stopped, that was it. We sent for the undertaker. (Cerveny, Balling and Cerveny 2022, 45)

One of the most notorious of the makeshift towns that grew up along the shore of the Colorado River was known as 'Ragtown.' Being so close to the river didn't help the inhabitants cope with the stifling heat.

> Helen Holmes: When you went down from the top of the hill down to the river, you could feel the heat. Sometimes you'd feel like you couldn't get your breath. It just seemed like it was so terrible hot . . . there were about 500 tents there over quite a large area.
>
> At night it was so hot that you had to set sheets to be able to rest because you just couldn't sleep, it was so hot. If there was a breeze and they would dry, you'd get up and wet them again in order to be able to sleep. (Cerveny, Balling and Cerveny 2022, 45)

How hot did it get?

According to Walker Young:

> I had temperatures taken myself that I'm positive were correct. When we were down at the damsite itself in Black Canyon, we did have temperatures recorded. . . . I went over in the shade and sat with my thermometer and watched the crew. It made a temperature of 140 degrees [°F, 60°C]. The foreman in charge sat with me. You know what his job was? He watched the men in his crew and when one of them staggered out, he told him to come on over and sit down here for a while. He'd sit there 15 minutes [and then go back to work.
>
> No verified records of such extreme temperature (140°F, 60°C) have been accepted by the US extremes committee or by me at the World Meteorological Organization. But the current official highest recorded global temperature of 134°F (56.7°C) was recorded only a few hundred miles away from Hoover Dam in Death Valley in 1913. (Cerveny, Balling and Cerveny 2022, 46)

Working in such extreme conditions took a horrific toll on the men and their families. Doctor Clare Woodbury was one of the physicians in Las Vegas who treated the workers overcome by the heat:

> At that time, very little [was] known about heat exhaustion. During the summer months, for a while they brought in sometimes 10 and 12 men at a time unconscious and incontinent. We had no electric thermometers then, but with an ordinary thermometer their rectal temperature went up to 112 degrees [°F, 44°C]. [At the beginning of construction] the mortality was about 50 percent or 60 percent. (Cerveny, Balling and Cerveny 2022, 46)

A government study by Dr. David Dill reported on one attempted fix to the heat:

> There was one lack that we recognized. The evidence had been advanced in the literature that workmen in high temperatures often have a deficiency of

> salt, develop heat cramps. . . . [The surgeon] had a big sign up in the mess hall: THE SURGEON SAYS DRINK PLENTY OF WATER. Dr. Talbott persuaded him to add to that: AND PUT PLENTY OF SALT ON YOUR FOOD. So that was the only step that we took to protect the men against loss of salt in their sweat. (Cerveny, Balling and Cerveny 2022, 46)

But the construction and maintenance of the Hoover Dam was vital—first to aid in the recovery from the Great Depression but more and more to preserve the critical power it produced for the blossoming Southwest.

Today that power generation stands in jeopardy, not because of construction, but because of drought. Unfortunately, the water levels of Lake Mead—the lake created by the Hoover Dam—in recent years has declined at some of its lowest levels since the 1930s, more than ninety years ago. As of 22 August 2021, Lake Mead was filled to just thirty-five percent of its capacity. That low water level comes at a time when ninety-five percent of the land in nine Western states have been affected by some level of drought. So, ending this interlude on a somewhat depressing note, after decades of humanity strenuously battling the forces of nature at the iconic Hoover Dam in the American Southwest, it might be that nature will have the final victory.

# 6 The World's Highest Recorded Wind Speed

You find out the strength of a wind by trying to walk against it, not by lying down.
C. S. Lewis

A few years ago, I was asked to write an encyclopedia article on the topic of winds. My first sentence for that entry was 'Wind is simply the movement of air.' Yes, wind is a rather basic weather concept. Air movement—or wind—originates from atmospheric pressure changes, the differences across the surface of our planet between high- and low-pressure systems. The larger the difference that exists between high- and low-pressure systems, the stronger the winds that blow between the two.

To visualize this concept, think of high pressure as a huge 'air mountain' and low pressure as a deep 'air valley.' The higher the high pressure, the taller the air mountain is, while the lower the low pressure, the deeper is the air valley. Mother Nature likes to equalize things and, in her attempt to equalize pressures, she uses wind. Wind can be thought as the air flow between the 'air mountain' and the 'air valley'. Air flows from high pressure towards low pressure. The greater the difference between high pressure and low pressure (or, using my analogy, the steeper the elevation difference between the air mountain and the air valley), the stronger are the winds.

In part, that is why the winds in hurricanes and other storms are so strong. The tropical cyclone's core is intense low pressure compared to the higher pressures existing away from the storm.

We measure wind speeds using an anemometer. The most common anemometer, still used across the globe today, is the hemispherical cup anemometer, which has three or four small hollow metal half-cups (hemispheres) set to catch the wind and revolve around a vertical rod. The number of cup revolutions over specific increments of time is used to calculate the velocity of the wind. Because wind speeds are not consistent—there are gusts and lulls—wind speed is usually averaged over various short periods of time.

Today, besides the cup anemometer, there are also other—more technical—instruments that calculate wind speed in several different, and strange, ways. For example, a hot-wire anemometer uses the concept that flowing air cools a heated object as it moves over it. In a hot-wire anemometer, an electrically

DOI: 10.4324/9781003367956-6

heated, thin wire is placed in the wind. The amount of power needed to keep the wire hot is key to calculating the wind speed. The higher the wind speed, the more power is required to keep the wire at a constant temperature.

Another type of wind-measurement instrument, a sonic anemometer, utilizes sound waves to calculate wind speed. It has no moving parts and relies instead on sonic pulse technology to measure both wind speed and direction. Today, there are even laser doppler anemometers. They emit a beam of light from a small laser. When the air molecules are in great motion, the light hitting them experiences a slight doppler shift in its wavelength that can be used in measuring wind speed. Laser doppler anemometers can measure very slight changes in airflow but they are very expensive.

Okay, we can measure wind speed. But what are the fastest surface winds ever recorded by humans?

When the WMO's Archive of Weather and Climate Extremes was created in 2007, the official accepted record for the highest recorded wind speed was 231 mph (103.3 m/s), recorded back in 1934 at the famous Mount Washington Observatory in New Hampshire, USA.

Sorry, I need to geek out for a moment or two to celebrate that most interesting and unique weather institution.

There are places around our planet that take great pride in holding a world record for weather extremes. Such is the case for the Mount Washington Observatory. Its slogan—'home of the world's worst weather'—is a point of great pride to the citizens of that state and to the many dedicated atmospheric scientists who have worked there.

Since 1932, Mount Washington Observatory has been a working, nonprofit scientific and educational institution. The weather observation station—still manned by a stalwart group of meteorologists, even in this day of automation—is located on the cold windy summit of Mount Washington. By maintaining its mountaintop weather station, the observatory's mission has been to advance our understanding of the earth's weather and climate by conducting research and education programs.

The observatory has a long and storied past.

Indeed, even before the observatory was created as a private scientific institution, the first regular meteorological observations on Mount Washington were conducted by the US Signal Service, a precursor of the Weather Bureau, from 1870 to 1892. Why Mount Washington?

The winds. The blistering hurricane-strength winds of Mount Washington are notorious around the world.

Part of the reason is that the air above the summit is one of the favored locations of something that we call a 'jet stream.'

As I mentioned in chapter 2, jet streams—fast flowing 'rivers' of air that blow from west to east and travel high in the atmosphere—were first discovered by the German meteorologist Heinrich Seilkopf in the first half of the twentieth century and were primary reasons for the extraordinary flight abilities of the German Zeppelin fleet. Although jet streams were recognized as real weather phenomena only in the 1930s, the unwitting measurements of some of their effects—like the accelerating winds blasting through mountain gaps in the Northeast United States—had been known for decades prior.

As far back as 1870, a group of scientists undertook an expedition to observe and record Mount Washington's brutal weather—and amazing winds. Over the course of their trek, those scientists gathered a wealth of information and captured the attention of the US Army Signal Service (the predecessor to today's National Weather Service). As a result, the Army Signal Service approved and then constructed a permanent weather station at the summit of Mount Washington. One of the first of its kind anywhere in the world, this mountaintop office remained operational until 1892.

Almost half a century later, a group of civilians founded Mount Washington Observatory. Using modest funds from research grants, they wanted to continue the work of the Signal Service. Their mission was to live atop the wind-blasted mountain, taking observations, and hoping to advance our understanding of climate and weather.

In 1934, they made history.

In that year, on the bitter cold morning of April 12, Wendell Stephenson, one of the observers at the weather station, woke at around four in the morning to the sound of extreme winds—some of the fastest he had ever heard. When he went to check the wind speed, Stephenson found the anemometer was clogged by ice, preventing the cups from properly spinning. With no other option, the meteorologist suited up to go outside, grabbing a wooden club on his way out the door. The wind was so powerful that he was knocked to the floor by the force of the gale the moment that the door opened. Step by step, he battled through the bitter wind to the anemometer, chipping away at the ice that had collected on the instrument. A moment after he finished, the strong winds tore the wooden club from his hand, carrying it off into the mountains and out of sight.

Returning to the station, the other observers told Stephenson that his hasty repairs had actually worked. They told him that the winds were slamming the observatory at 150 miles per hour and climbing fast.

As that stormy day back in 1934 continued, the winds grew more powerful. Then, a bit after 1 pm in the early afternoon, the instrument recorded a record-shattering wind speed of 231 mph (103.3 m/s). After the storm had passed, the anemometer was immediately subjected to a series of several calibration tests and the National Weather Service announced that the observatory's reading was an accurate and valid reading.

In the 1990s, one of my former students had the extraordinary opportunity to interview one of those legendary observers of the 1930s, Mr. Alex McKenzie. McKenzie had been one of the observers who had recorded the 231-mph (103.3 m/s) wind speed. In his eighties, the meteorologist relayed to my student the many difficulties of working at the observatory in those early days. McKenzie recounted the story when, once back in 1936, Mount Washington's savage winds ripped the roof of the Observatory completely *off* of the building! After the observatory was repaired, the meteorology crew used a series of strong steel chains strung across the roof to hold the structure in place. And, yes, those chains still remain to this day.

*Figure 6.1* Left, Southern Illinois geography professor Dr. Mark Hildebrandt standing in front of a weather display at the Mount Washington Observatory; right, the plaque commemorating the 'highest wind ever observed by man' at the Mount Washington Observatory.
Source: Photographs courtesy of Hildebrandt.

Today, the Mount Washington Observatory is a private, nonprofit organization dedicated to climate and weather research. That devotion is clearly evident in both the actions and words of my former student, whom I have mentioned above. Dr. Mark Hildebrandt, now a professor of geography at Southern Illinois University, had the fortunate opportunity to serve at the observatory. He kindly recounted to me some of his own past work:

> I worked at the Mount Washington Observatory during the winter of 1992–1993. Each shift would last for one week, with Wednesdays being shift change day. As an intern, my duties included performing weather observations every three hours and reporting each observation to the National Weather Service office in Gorham, Maine. Among my other duties was maintaining and de-icing the instruments in the harshest conditions. I recall carefully knocking rime ice off of the wind vane in hurricane-force winds in temperatures well below zero at 3 AM.
>
> I was on the summit during the 'Storm of the Century' in March 1993. The winds averaged over 80 mph for days, and I recall seeing snow drifts inside the building as the wind had blown snow through the caulking in the windows.
>
> The strongest sustained wind I experienced was on April 24, 1993. Sustained winds were in excess of 100 mph with gusts over 140 mph. During this event, I tried walking on the roof of the Observatory. However, the wind was so strong that it knocked me off of my feet and it blew me about 30 feet across the roof. I had to crawl on my hands and knees to get back

> inside; it was a case of three crawls forward, two crawls back. This struggle lasted for a few minutes, though it seemed like an hour. I remember the force of the wind being so strong that breathing was very hard as the wind made it difficult for my chest to expand. Once I was inside, the wind sounded like a revving jet engine had been parked outside.

And for the next sixty years, following that observation of 231 mph (103.3 m/s) in 1934, Mount Washington Observatory maintained its record of having recorded the world's fastest wind speed.

Then, in early April of 1996, a huge tropical cyclone named Olivia hit the northwest coast of Australia in the Southern Hemisphere. The tropical storm developed north of the city of Darwin and rapidly intensified into cyclone (hurricane) strength. By 9 April, it had reached Category 4 wind speeds as measured in the US—although the US and Australian tropical cyclone ranking systems are slightly different, both indicated similar intensity. A Category 4 classification would translate to wind speeds of 105 knots—fifty-four meters per second, or over 120 miles per hour.

At that peak intensity, the eye of tropical cyclone Olivia passed by a small island named Barrow Island. That island is a 202-square-kilometer (78 sq. mi.) island, fifty kilometers (31 mi.) northwest off the Pilbara coast of Western Australia. Olivia made landfall near Mardie station, west of Karratha. Mardie and other surrounding stations suffered extensive damage to their properties. Several sites recorded extreme wind gusts during Olivia. Varanus Island reported a wind gust of 74 m/s (166 mph), and the Mardie station registered a wind gust of 72 m/s (161 mph).

Barrow Island recorded the highest wind speeds. On 10 April 1996, as tropical cyclone Olivia passed close by, a privately operated automatic anemometer on that island measured a wind gust of 113.3 m/s (253.4 mph).

Those Barrow Island winds were registered by an automatic weather station using a heavy-duty three-cup anemometer. The wind observations made by the Barrow Island anemometer were recorded every five minutes, with the mean wind being the five-minute average and the strongest gust being the peak three-second average gust over the five-minute observation. The peak wind gust of 113 m/s (252 mph) was one of a series of extreme gusts during a series of five-minute time periods. During Olivia's passage, wind gusts of 102, 113 and 104 m/s (228, 252 and 233 mph) were measured. That middle one was the one in which we were interested.

But was it a legitimate observation? The WMO was brought into the case. I put together a committee of expert meteorologists, including atmospheric scientists from Australia, France and Great Britain and two renowned hurricane experts from Cuba and the US Hurricane Center. For the committee's later scientific research paper on the subject, we were pleased to have the Director of the Mount Washington Observatory join us as well.

It was a long and detailed examination. My evaluation committee had many questions. But one fact in particular was a key to this investigation.

Even after experiencing such extraordinary hurricane-strength winds, the Barrow Island anemometer itself *survived*. As one committee member noted, 'I am surprised that the instrument itself and the supporting mast survived these extreme conditions—a tribute both to the designers of the anemometer and those who installed it.'

Other components of the Barrow Island weather station suffered damage. For example, the radiation shield on the temperature sensor was damaged and heavy rains infiltrating into that sensor may account for a few unrealistic temperatures. It is also likely that the relative humidity sensor was wet with rain causing one hundred percent humidity during the period of strong winds. But the committee concluded that in no way did those failures invalidate the wind data, as the temperature and relative humidity probes were independent of the anemometer.

Some of the arguments discussed within the evaluation committee were a bit technical. For example, one committee member wrote that:

> [T]he very radical changes in the gust factors . . . suggests to me there was an instrumentation problem. Now, the documentation states that the instrument data logger had some safeguards to prevent false reading. I'm wondering, though, if there was some failure mechanism that was not accounted for by these safeguards? What could cause false readings in this instrument in these circumstances?

In response, other members (and an instrument consultant who we brought into the discussion) noted that, if the instrument had been operating with speeds at ninety to one hundred knots (46.3–54.4 m/s; 103.6–115.1mph), with gusts to 150 knots, and spiked to over 200 knots *with nothing else even near that*, then it would raise major red flags. However, the fact was that the Barrow Island anemometer recorded *three* gusts at or near 200 knots range, not just one, lends considerable confidence to the extreme being legitimate.

*Figure 6.2* The Barrow Island (Australia) anemometer.
Source: Photo courtesy of MetOcean Engineers.

Another committee member opined that:

> [O]f course one can never be 100 percent sure that an instrument is reading correctly, but there is no reason, in principle, why the anemometer in question should have been overestimating peak gusts. . . Cup anemometers will overestimate the mean wind in gusty conditions (as they respond more rapidly to an acceleration in wind speed than a deceleration), but the peak gusts should be accurately recorded. Wind tunnel calibrations of a similar instrument confirmed its accuracy at these high wind speeds.

That was another key experiment discussed by the evaluation committee. An identical anemometer to the Barrow Island instrument was thoroughly tested in a NASA high speed wind tunnel, which confirmed a proper wind speed response for the instrument. Such studies have given greater credence to the validity of the Barrow Island observations.

Finally, a couple of committee members voiced some concerns about the nature of the event—the gust having occurred within a tropical cyclone. They wondered, even though the winds of a tropical cyclone can be fierce, how could such tremendous winds—a wind gust of 113.3 m/s (253.4 mph)—occur? Remember that the cyclone itself had been ranked having wind speeds of 105 knots, fifty-four meters per second, or over 120 miles per hour. This observation was well above that ranking.

A majority of the committee agreed that it was likely the immense wind extremes were the result of concentrated strong winds *within* the tropical cyclone, something like a small, embedded tornado, or what we call a 'meso-vortex.' Other studies have shown that such small intense turbulences can occur with tropical cyclones. For example, published research has indicated that several tornadoes and downbursts occurred in 1992's horrific Hurricane Andrew, as it made landfall in Florida. What this means—of great importance to planners and engineers—is that they should be aware that extreme gusts, which are much higher than the typical gusts associated with the tropical cyclone naming system, can occur—particularly for strong tropical cyclones.

After a long, detailed, and (for a meteorologist such as myself) very interesting investigation, the committee recommended that the WMO should recognize the observation as legitimate. I accepted their recommendation and in 2009, the WMO officially accepted the Barrow Island wind extreme of 113.3 m/s (253.4 mph) as the highest recorded wind gust for the world.

Yet that announcement led to some interesting reactions from the public and media. One somewhat confusing aspect to the public was the long lag time between occurrence of the wind event (April 1996) and the deliberations of the WMO evaluation committee (2009). Several factors played a role in that delay.

At the time of Cyclone Olivia's passage, the extreme wind measurements at Barrow Island were initially viewed with skepticism and were not included in the post analysis report conducted by Australia's Bureau of Meteorology (BoM). This was in spite of the fact that observations from nearby Varanus

Island were included and given status as the strongest gust ever recorded in Australia. Part of this absence of mention in the BoM report was because the Barrow Island instrument was not owned and operated by the Bureau of Meteorology; it was privately owned by an energy company. Unfortunately, the wind data had not been incorporated into the official Australian observational database and that contributed to the lack of a follow-up investigation.

Consequently, the WMO never heard about it back in 1996. I should note that one of our committee members in particular, Blair Trewin—a superb climatologist of Australia's Bureau of Meteorology—played a major role in bringing this particular observation to the WMO's notice. Blair is a gifted Australian climate scientist whose specialties are the development of long-term historical data sets for the assessment of climate change, and the analysis of extreme events, both current and historic. To give you an idea of the regard in which Dr. Trewin is held down under, he has been the President of the Australian Meteorological and Oceanographic Society. Dr. Trewin is an incredible climatologist who is fascinated with weather extremes—and that interest led to the WMO's investigation into Tropical Cyclone Olivia's winds.

Then, the second reaction to our WMO announcement was a bit more on target.

Several of the media told me, 'Wait a moment! The fastest winds of the world? *Only* 253 mph?'

And they are right. Without question, there are stronger winds on our planet than 253 mph. But our announcement was for recorded winds.

We know the winds of the strongest tornadoes, for example, can be significantly stronger than that (see chapter 11). Why aren't those winds regarded as the fastest observed winds?

At this time, we at the WMO Archive of World Weather and Climate Extremes only accept measured wind speeds—not estimated—as the official highest recorded winds. For example, although the 1999 Moore Oklahoma tornado (with an F-5 ranking on the Fujita scale) produced winds estimated by Doppler Radar at over 300 mph (134 m/s), those winds were estimated—not directly measured by an anemometer. Therefore, the Barrow Island observation of 113.3 m/s (253.4 mph) remains the highest *recorded* wind speed.

On a personal note, this evaluation was bittersweet. Although—as was becoming common—my WMO panel of experts had performed admirably in assessing the validity of Tropical Cyclone Olivia's incredible winds at Barrow Island, the new record meant that the Mount Washington Observatory's record of 231 mph had been superseded. That legendary weather institution in New Hampshire no longer held the world extreme of highest observed wind speed.

And my melancholic feeling was echoed by many of the residents of the state of New Hampshire where the observatory resides. During an interview with me about the new record, one New Hampshire reporter sadly noted that the state only had two major claims to fame. First, he said, one recognition for the state had always been the 'Old Man of the Mountain,' a prominent rock formation displaying a face-like feature. And the second distinction for New Hampshire was 'the Big Wind,' the extreme wind measured at the Mount Washington Observatory.

With a tinge of true sadness, the reporter then told me that the 'Old Man of the Mountain' formation had unexpectedly collapsed in 2003, and 'now you're taking the Big Wind away from us. What's left for us?'

In reply to all Granite Staters: I do like your state, and let me be quick to point out that we at the WMO continue to monitor all weather extremes. So, if at some time in the future, your valiant meteorologists at the Mount Washington Observatory do measure a stronger wind gust, I would be most pleased to open another investigation to verify it!

# Interlude: Freaks of the Wind

Luckily, as I wrote this chapter on our WMO wind evaluation, I did *not* suffer the strange problem that Camille Flammarion experienced back in 1872 in Paris, France.

In the nineteenth century, the great Flammarion was a 'rock star scientist'—a nineteenth-century Carl Sagan or Stephen Hawking. Like those two modern scientists, Flammarion's brilliance had captured the general public's interest and admiration. Part of that esteem was due to the popularity of several books that he had written on meteorology and astronomy. And that's where this particular 'weather freak' event begins.

As he was writing one of his weather tomes, Flammarion experienced a strange wind-related incident. In 1900, the great scientist wrote how, back in 1872 when he was finishing the chapter on wind in his early landmark work on the atmosphere (*L'Atmosphère*), a blustery gale blew open his window, lifted the loose pages he had just written, and carried them off:

> During the time I was writing my great book on the atmosphere, I was busy with the chapter on the forces of the wind, and I was comparing several curious examples when the following thing took place:
>
> My study in Paris is lighted by three windows, one looks east on the Avenue de l'Observatoire, another southeast towards the Observatory, the third to the south on to the Rue Cassini. It was the middle of summer. The first window was open, looked on the chestnut trees of the avenue. The sky was clouded; the wind rose, and suddenly the third window, which must have been badly fastened, was violently blown open by a gale from the southwest, which disarranged all my papers, and lifting the loose pages I had just written, carried them off in a sort of whirlwind among the trees. A moment after the rain came, a regular downpour.
>
> To go down and hunt for my pages would seem to be to be time lost, and I was very sorry to lose them.
>
> What was my surprise to receive, a few days later, from Lahure's printing-office, in the Rue de Fleurus, about half a mile away from where I lived, that very chapter printed without one page missing.
>
> Remember, it was a chapter on the strange doings of the wind.

> What had happened?
>
> A very simple thing.
>
> The porter of the printing-office, who lived at the Observatory, and who brought me my proof-sheets as he went to breakfast, when going back to his office noticed on the ground, sodden by the rain, the leaves of my manuscript. He thought he must have dropped them himself, and he hastened to pick them up, and having arranged them with great care, he took them to the printing-office, telling no one of the affair.
>
> A little more, and some credulous person might have asserted that it was the wind that had brought them to the printing-office. (Flammarion 1901, 192)

Such a lucky occurrence as Flammarion experienced might be called a quite fortunate 'windfall.' Merriam-Webster gives a definition of *windfall* as 'an unexpected, unearned, or sudden gain or advantage.' But the formal usage of the word dates to 1362 to a famous windstorm that struck the British Isles.

A scribe, writing at that time in the *Chronicon Angliae*, wrote that:

> A vehement wind burst forth with such strength that its blowing violently threw down high houses, tall buildings, towers, bell turrets, trees and other strong and durable things and likewise it raged in such manner that the remains which are still extant are weaker even until now. (Anonymous 1930, 73)

Damage from the violent storm included the utter destruction of bell towers in Bury St. Edmunds, Norwich and London.

In particular, the so-called Black Prince of England (Edward, the eldest son of Edward III) lost a considerable amount of timber across his vast estate.

Across England at that time, a tenant could not cut trees on his land. By royal decree, all *living* timber belonged to the king. Cut wood from those lands was only to be used in building the Royal Navy's great ships. *But* an estate owner could keep any trees felled by the wind, which was considered a divine act. Such trees could be sold for profit, thus causing the common class to view such 'windfalls' as being most favorable events. According to chroniclers of the time, the Black Prince lost a considerable amount of timber due to the terrible storm of 1362—lumber at was immediately picked up, used and sold by the locals holding those lands.

And so the term *windfall* was born.

As a writer in an 1848 issue of *American Scientist* wrote:

> Some of the English nobility were forbidden felling any of the trees in their forests—the timber being reserved for the use of the Royal Navy. Such trees as fell without cutting were the property of the occupant. A [great wind] was, therefore, a perfect Godsend, in every sense of the word, to those who had occupancy of these extensive forests and a windfall was sometimes of very great value. (Anonymous 1848, 113)

But sometimes strong winds can aid a king or an emperor. An excellent example of this relates to a battle of Theodosius the Great, the last emperor to rule the entire Roman Empire. In 394 CE, Theodosius was under deadly attack by a rival named Flavius Eugenius. After diligent praying for divine aid on the eve of the engagement, Theodosius took to the battlefield. As the two forces came within fighting distance, a powerful, immense windstorm began to blow. Specifically, the blustery wind stormed onto the battlefield from *behind* Theodosius's forces and blew *towards* the faces of the enemy.

Therefore, the javelins of Theodosius's army flew high through the air and plunged deep into the heart of the enemy's forces, causing great death and destruction. Conversely, the weapons of the opposing army, even though hurled with all the warriors' might, were caught by the wind and—according to the chroniclers—even tossed backward. Their own javelins were blown backward to kill and maul Eugenius's enemy throwers. Unsurprisingly, shortly thereafter, Eugenius was captured and killed.

Emperor Theodosius the Great was victorious . . . thanks to the powerful winds of a divine storm!

# 7 The Most Rain, Normals, Misplaced Decimals and Other Errors

. . . I always set down positively what weather my reader will have, be he where he will at the time. We modestly desire only the favorable allowance of a day or two before, and a day or two after the precise day against which the weather is set, and if it does not come to pass accordingly, let the fault be laid upon the printer, who, 'tis very like, may have transpos'd or misplac'd it, perhaps for the conveniency of putting in his holidays: and since, in spite of all I can say, people will give him great part of the credit of making my Almanacks, 'tis but reasonable he should take some share of the blame.

Benjamin Franklin, writing as 'Poor Richard' on taking responsibility for his almanack forecasts

The amount of water that can fall from the sky can be awe inspiring.

Recording the sometimes-enormous quantities of falling water can be daunting.

Most of us are familiar with the standard rain gauge. It is a basic cylindrical container that is open at the top that allows rainwater to enter. One would think that such a simple device has been around for centuries, perhaps millennia. But, beyond a few isolated outliers mentioned in ancient India, Biblical Palestine, and Southeast Asia, no scientific records of rainfall measurement exist until the seventeenth century.

The originator of systematic recording of rain amounts?

Benedetto Castelli, a Benedictine monk and student of Galileo (of thermometer fame as mentioned in chapter 3) is credited by most historians with the invention of the rain gauge. He wrote to the famous Italian scientist, Galileo, in 1639:

> I took a glasse formed like a cylinder, about a palme high, and half a palme broad [or about four inches in height and two inches in diameter]; and having put in it water sufficient to cover the bottom of the glasse, I noted diligently the mark of the height of the water in the glasse, and afterwards exposed [it] to open weather, to receive the rain water, which fell into it; and I let it stand for the space of an hour; and having observed that in that time the water was risen in the vessel the height of the following line [about a third of an inch long to represent the depth]. I considered that if I

DOI: 10.4324/9781003367956-7

> had exposed the same rain such other vessel equal to that, the water would have risen in them according to that measure. (Gedes 1930, 4)

Today, a more sophisticated, but still simple, top-opened container is a common instrument for recording rainfall. No specialized training is necessary for operation. Anybody can set up and operate such a rain gauge.

For example, since 1998 a huge nationwide effort now exists across the United States to collect rainfall records using ordinary citizens as observers. Called the Community Collaborative Rain, Hail and Snow Network (*CoCoRaHS*), the volunteer project is a nonprofit, community-based network of people of all ages and backgrounds that is designed to measure and map precipitation across the entire country. The data collected by these citizen weather enthusiasts are then used by many organizations and individuals, such as the National Weather Service, hydrologists, emergency managers, city utilities, insurance adjusters, USDA, engineers, mosquito control, ranchers and farmers. In particular, many teachers use the CoCoRaHS rainfall data in student course work. The precipitation information advances our ability to understand weather, and definitely helps to promote meteorology.

Interested in becoming a citizen meteorologist? You can check out the CoCoRaHS website and learn the details of the project at https://www.cocorahs.org/.

One question that arises with rain measurement is, Does the size of the opening atop a standard rain gauge matter? Actually, no. Testing has shown that accurate rainfall collection doesn't depend on the size of the rain gauge's mouth, if the opening is at least four inches in diameter.

Although the cylinder-based rain gauges of CoCoRaHS are effective, easy-to-use and relatively cheap (costing about $40 US), they don't provide minute-by-minute accurate recording of rainfall. For that, most meteorological weather stations use tipping bucket rain gauges.

Tipping-bucket rain gauges date back to the early 1800s. In 1829, a scientist named Crossley designed the first double-sided tipping bucket rain gauge. Rain falls into a large opening (thirty-centimeter, twelve-inch) and is funneled down to a small spoon-like paddle. When a specified amount of water, usually around 0.25 millimeters (0.01 inch), falls onto that paddle, it overloads the paddle, which falls and empties out. When that action happens, another paddle on the other side is moved up and positioned under the funnel. Each tip of a paddle is recorded. The tips are logged both as an accumulative number (total storm amount) or by the time and date of each tip.

For the most part, the device operates problem free, although some difficulties still occur. For example, if a massive amount of rain occurs in a very short amount of a time, the tipping spoons can't keep up and water will back up in the funnel. They will still accurately record the total amount of rainfall, but the timing of specific amounts might be affected due to the time for all the water to work through the funnel.

A bigger concern with rainfall measurements—and with all weather-related observations—is often human error.

For instance, I discovered a published technical report that stated that in 2003 the remarkably wet village of Cherrapunji, India recorded a twenty-four-hour rainfall total of 1,840 mm—a bit over six feet of rain in one day!

Cherrapunji, India is legendary in the annals of meteorology. Its fame is that, a long time ago, in the nineteenth century, over the course of twelve months (August 1860 to July 1861), the weather observer stationed at that location recorded an astounding 26.47 *meters* of rain (1042 inches or nearly 87 feet of water). That is *still* the world-record twelve-month rainfall amount. Whenever we at the WMO see Cherrapunji mentioned in an official record, we tend to take notice!

So, when I saw a report that in 2003 Cherrapunji, India recorded a twenty-four-hour rainfall total of 1840 mm (72.4 inches), I quickly started an initial investigation. Such a high rainfall value would be comparable for our existing twenty-four-hour rainfall record. Currently, the twenty-four-hour world record for rainfall is associated with the passage of a tropical cyclone that, in 1966, dropped an astounding 1.825m (or 71.8 inches) of rain in just twenty-four hours, on the isolated French mountainous and volcanic island of Réunion, in the South Indian Ocean.

Regarding Cherrapunji, I contacted the efficient India Meteorological Department. In turn, they immediately launched an internal investigation. Within a matter of mere days, they had identified a problem with that record rainfall.

Those Indian meteorologists went back to the very important original Cherrapunji log sheets for 2003 (remember the critical log sheets discussed in our Libyan temperature investigation chapters). When they did so, they found that the reported 1840 mm value was actually an inadvertent typographical error.

They determined that the true rainfall at Cherrapunji on that particular day in 2003 was only 184 mm, not 1,840 mm. A decimal point had been misplaced. Only a misplaced decimal point? Such an error is the difference between being a world record . . . or not. While a rainfall of 184 millimeters is still a respectable amount—a bit over seven inches of rain in twenty-four hours, it is nowhere close to a new world record.

So, errors can sometimes creep into the world's weather records. One of my tasks as world weather judge is to make sure—as much as I can—that those errors don't get into the Archive of World Weather and Climate Extremes.

One strange error that can sometimes creep into our weather records involves units.

The importance of the precise use of measurement units in science was evident back in 1999 when the US Mars Climate Orbiter probe was lost. It had been attempting an orbit insertion around the red planet—and it failed. After an investigation, NASA officials determined the loss of the probe was due to a misunderstanding over which units—English or metric—were being used to fine-tune the spacecraft's trajectory.

In essence, a failure in communication occurred between the two groups in charge of the probe. The spacecraft team at Lockheed Martin Astronautics in Denver, which determined how the satellite was to fire its thrusters, was using English units in its measurements, a practice still common in some engineering circles, such as defense contracting. In contrast, the Jet Propulsion Laboratory navigation team for the orbiter—the group which determined the spacecraft's position and how its trajectory should be changed—was using metric units. When JPL group received the Lockheed Martin data, they had assumed the data were in metric units, which was the convention agreed to at the outset of the mission.

The engine fired precisely as based on the inputted data but, because of the conversion error, the spacecraft came within 60 km (36 miles) of the planet—about 100 km (60 miles) closer than planned and about 25 km (15 miles) *beneath* the level at which the engine could function. That extremely close approach to Mars caused the spacecraft's propulsion system to overheat and prevented the engine from completing its essential burn. The Climate Orbiter ploughed through the thin atmosphere, continued out beyond Mars and is now aimlessly orbiting the sun.

Units are important.

In meteorology, repeated conversion between English and metric units can introduce error. I've found a couple times in the past academic literature that some meteorologists have converted records originally written in metric units into English units, which a later researcher then reconverted back into metric. How can that lead to error?

Let's suppose, for example, that someone recorded a rainfall observation of 4.23 inches of rain.

Using the conversion between inches and millimeters, one could translate that value into a precise rainfall of 107.442 millimeters. But suppose that translated value was rounded to, and published as, 107 mm. Then, later, another researcher sees that converted rainfall amount and reconverts it back into English units. The result? They get a rainfall value of 4.21 inches instead of the actual 4.23 inches.

That may not seem like a huge difference but when we are talking about scientific accuracy of weather extremes and thousands of records, it can be quite important. And if the back-and-forth conversions are repeated, the error can grow. Repeated conversions are bad.

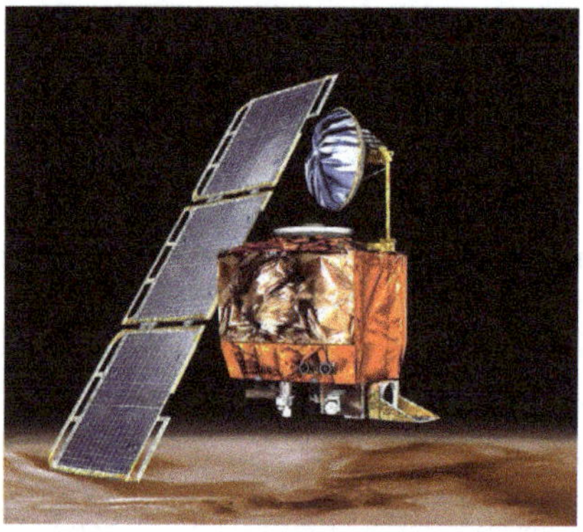

*Figure 7.1* An artist's concept of the failed NASA Mars Climate Orbiter.
Source: Courtesy of NASA/JPL-Caltech.

And that is just one problem.

Rain, in particular, is a weather variable that seems to attract a wealth of strange difficulties.

An interesting case of one of those problems involves a record category that most people would consider easy to determine. That would be determination of the 'wettest place on earth' (e.g., the location that receives the greatest amount of rainfall over the course of a year).

One only needs to look at the world's rainfall records and see where it rains most, right?

We have been unable (so far) to determine a clear candidate for that record. The issue revolves around the complex concept of a climatic *normal*.

The idea of a climate normal dates to that pioneering 1872 International Meteorological Congress in Vienna that I mentioned in chapter 2. One of the resolutions of those innovative meteorologists was to compile for all weather stations across the world the weather values averaged over a uniform, multiyear time interval. This would allow them to have a comparison between climate data collected at various stations. In other words, they wanted to establish a long-term average for each weather station around the world. They called such a long-term average a 'climate normal,' after usage first employed by a meteorologist named Heinrich Dove in the 1840s.

Those Vienna scientists believed that a climate normal would serve two critical purposes. First, it could serve as a benchmark—a baseline—against which recent or current observations can be compared. For example, in today's world, most evening television meteorologists when they discuss the given day's high and low temperatures will also mention the normal high and low temperature for that date. That normal (which is based on these early meteorologists' work) would be a reference average temperature to help the viewer see if today's values were higher or lower than those recorded from the past.

The second purpose that a climate normal serves is to give a prediction of what a person might expect to happen weatherwise on a given day. For example, if someone was planning a vacation to an exotic location, that area's climate normals would give them an idea of what weather conditions they might expect.

Okay, the concept of a climate normal is a pretty good idea. But for meteorologists and climatologists, a key concern is what *specific* length of time should we use in calculating such a long-term climate average. A ten-year average? Twenty? Thirty? Fifty years or more? Part of the problem was when the idea of *normals* was first globally accepted back in the 1930s, many weather stations didn't have long periods of record while many others had been around for many decades.

The issue wasn't an easy one. Only after decades of extensive discussion was an international compromise reached to define a climate normal as a thirty-year period. Beginning in 1935, the international forum of meteorologists around the world decreed that the period for a climatic normal would be from 1901 to 1930.

Since that time, the WMO has made several changes to the definition of a climate normal. The most significant of those changes was a sliding thirty-year base. A climate normal now refers to the *most-recent* thirty-year period finishing in a year ending with *0*. Currently (at the time of this writing), the WMO's thirty-year normal is 1991–2020. Into the 2030s, we will switch to climate normals for the time interval from 2001 to 2030.

If someone says 'above normal precipitation' or 'above normal temperatures,' they are referring the average of the last official thirty-year period. That is consistent across the world. Like the measurement of weather in general, we want a normal to be the same around the entire world. Russia's normals should be the same as those of the US or Australia's. So, the WMO requires its members to compute their individual country's climate normals using those criteria on what years to use and how long that period should be.

But how does that cause a problem in determining the wettest place on earth (e.g., the location that annually receives the greatest amount of rainfall)?

Several times in the past, I have formed evaluation committees to examine potential candidates for that record of the highest annually averaged precipitation, one of the most recent being the remote location of Puerto Lopez, Colombia. And, for that committee, I must in particular commend the work of my regional representative, Ruth Leonor Correa of Colombia's Instituto de Hidrología, Meteorología y Estudios Ambientales.

At first glance, Puerto Lopez appeared to be a good candidate. Puerto Lopez, Colombia is a very wet place as the result of being within the Intertropical Convergence Zone or ITCZ. The ITCZ is the region near the equator, where the moist trade winds of the Northern and Southern Hemispheres come together. That convergence—combined with lots of heat—warms the moist air and causes it to rise. As the air rises, it expands and cools, releasing the accumulated moisture in an almost perpetual series of thunderstorms. Equatorial countries like Columbia are always good candidates for rainfall records.

But the 'wettest place on Earth' record involves an *average* rainfall, not that of any specific year. The WMO specifies, the location must have at least thirty years of data—a climatic normal. We needed to look at the detailed records.

As we did so, one issue that my committee also had to examine was how continuous did that thirty years of data have to be. Could there be occasional missing days in that thirty-year period? If so, how many? Those questions and others made the committee's analysis quite involved. We had to delve into the many rules and regulations of the WMO regarding climatic normals. I didn't think when I started this project that the talents of a lawyer would ever be useful in extremes work, but such skills did seem to be the case as we addressed the WMO's many and somewhat complicated rules.

My evaluation committee evaluated Puerto Lopez's detailed rainfall data for the climatic normal of 1981 to 2010 (the period then under question). We compared that record to the WMO's existing rules and regulations. When we did so, we were only able to identify twelve complete years of rainfall records, not the thirty required. I couldn't accept Puerto Lopez as the wettest place on Earth.

So, I and the WMO announced that we were 'leaving the category of "highest annually averaged precipitation extreme" vacant at this time.' But we did add an important caveat: 'If more data from Puerto Lopez or another claimant for this extreme is brought to the attention of the WMO, a new evaluation committee may be established' (WMO 2014).

Finally, one other issue that occurs with weather observations—particularly precipitation—involves time.

In chapter 3, I mentioned that one concern in our evaluation of the Western Hemisphere twenty-four-hour rainfall extreme was what is the correct twenty-four-hour measurement period for the record? In meteorology, we use the world-accepted Coordinated Universal Time (UTC)—in essence Greenwich Mean Time—for many of the standard weather observations taken around the world.

For example, no matter where you are around the world, weather balloon (radiosonde) launches are *always* conducted at two specific times of the day: 00 UTC and 12 UTC. That means weather balloons are launched in London at midnight and noon local time, while—at the exact same time—they are being launched in New York at 7 PM and 7 AM local time and in Los Angeles at 4 PM and 4 AM local time.

Such uniformity means we always ensure that we are comparing apples and apples—in other words, the times we are using correspond to the times for the rest of the world. That idea can become complicated even with something as simple as the recording of a date.

In the United States, it is common to list a given calendar day as *month/day/year*. However, the rest of the world generally uses a *day/month/year* format. That can cause problems. When we at the WMO are working with a potential record that, for example, might be written as '4/1/2024,' we must always double-check to see if it is referring to the first day of April in 2024 or the fourth day of January in 2024!

If you check our WMO World Archive of Weather and Climate Extremes, you will see that we use the international *day/month/year* standard formatting—but with double-month notation (number and name) so that American viewers will hopefully not make a mistake, such as '4/ 1 (January)/2024'!

The take-away message: one always must make sure that one's *i*'s are dotted and *t*'s are crossed. Details matter in meteorology.

# Interlude: Freaks of Rain Magic

The legendary Dr. David Livingstone was a Scottish physician, anthropologist and missionary who is regarded as one of the great explorers of Africa during the nineteenth-century Victorian era. Livingstone's fame as an explorer and his obsession with learning the sources of the Nile River was founded on the belief that if he could solve that age-old mystery, the fame he would gain as a result could help him end the Swahili slave trade. 'The Nile sources,' he told a friend, 'are valuable only as a means of opening my mouth with power among men. It is this power [with] which I hope to remedy an immense evil' (Livingston 2006).

Beyond his missionary and exploration work, he also was an anthropologist who was intrigued with the beliefs and culture of the African people.

One specific interest of Livingstone involved 'weather magic' and, in particular, the witchcraft of rainmaking.

In his 1857 book, *Missionary Travels and Researches in South Africa*, Dr. Livingstone detailed a discussion of a long four-year drought where he addressed his African hosts' beliefs in rainmaking. As a good Christian, he compared those beliefs to his own nineteenth-century Christian dogmas. Although likely not Livingstone's intention, the following section gives a fascinating look at *both* sets of beliefs:

> But in our second year again no rain fell. In the third the same extraordinary drought followed. Indeed, not ten inches of water fell during these two years, and the Kolobeng [River] ran dry; so many fish were killed that the hyaenas from the whole country round collected to the feast and were unable to finish the putrid masses. A large old alligator, which had never been known to commit any depredations, was found left high and dry in the mud among the victims. The fourth year was equally unpropitious, the fall of rain being insufficient to bring the grain to maturity.
>
> . . . Rain, however, would not fall. The Bakwains believed that I had bound [the rain doctor] Sechele with some magic spell, and I received deputations, in the evenings, of the old counselors, entreating me to allow him to make only a few showers: 'The corn will die if you refuse, and we shall become scattered. Only let him make rain this once, and we shall all, men, women, and children, come to the school, and sing and pray as long as you please.

It was in vain to protest that I wished Sechele to act just according to his own ideas of what was right, as he found the law laid down in the Bible, and it was distressing to appear hard-hearted to them. The clouds often collected promisingly over us, and rolling thunder seemed to portend refreshing showers, but next morning the sun would rise in a clear, cloudless sky.

The natives, finding it irksome to sit and wait helplessly until God gives them rain from heaven, entertained the more comfortable idea that they could help themselves by a variety of preparations, such as charcoal made of burned bats, inspissated renal deposit of the mountain cony—'Hyrax capensis', the internal parts of different animals—as jackals' livers, baboons' and lions' hearts, and hairy calculi from the bowels of old cows—serpents' skins and vertebrae, and every kind of tuber, bulb, root, and plant to be found in the country.

Although you disbelieve their efficacy in charming the clouds to pour out their refreshing treasures, yet, conscious that civility is useful everywhere, you kindly state that you think they are mistaken as to their power. The rain-doctor selects a particular bulbous root, pounds it, and administers a cold infusion to a sheep, which in five minutes afterward expires in convulsions. Part of the same bulb is converted into smoke, and ascends toward the sky; rain follows in a day or two. The inference is obvious. Were we as much harassed by droughts, the logic would be irresistible in England in 1857.

As the Bakwains believed that there must be some connection between the presence of 'God's Word' in their town and these successive and distressing droughts, they looked with no good will at the church bell, but still they invariably treated us with kindness and respect. I am not aware of ever having had an enemy in the tribe. The only avowed cause of dislike was expressed by a very influential and sensible man, the uncle of [the rain doctor] Sechele. 'We like you as well as if you had been born among us; you are the only white man we can become familiar with; but we wish you to give up that everlasting preaching and praying; we cannot become familiar with that at all. You see we never get rain, while those tribes who never pray as we do obtain abundance.'

As for the rain-makers, they carried the sympathies of the people along with them, and not without reason. With the following arguments they were all acquainted, and in order to understand their force, we must place ourselves in their position, and believe, as they do, that all medicines act by a mysterious charm. The term for cure may be translated 'charm' ('alaha').

MEDICAL DOCTOR: Hail, friend! How very many medicines you have about you this morning! Why, you have every medicine in the country here.

RAIN DOCTOR: Very true, my friend; and I ought; for the whole country needs the rain which I am making.

MEDICAL DOCTOR: So you really believe that you can command the clouds? I think that can be done by God alone.

RAIN DOCTOR: We both believe the very same thing. It is God that makes the rain, but I pray to him by means of these medicines, and, the rain coming, of course it is then mine. It was I who made it for the Bakwains for many years, when they were at Shokuane; through my wisdom, too, their women became fat and shining. Ask them; they will tell you the same as I do.

MEDICAL DOCTOR: But we are distinctly told in the parting words of our Savior that we can pray to God acceptably in his name alone, and not by means of medicines.

RAIN DOCTOR: Truly! but God told us differently. He made black men first, and did not love us as he did the white men. He made you beautiful, and gave you clothing, and guns, and gunpowder, and horses, and wagons, and many other things about which we know nothing. But toward us he had no heart. He gave us nothing except the assegai, and cattle, and rain-making; and he did not give us hearts like yours. We never love each other. Other tribes place medicines about our country to prevent the rain, so that we may be dispersed by hunger, and go to them, and augment their power. We must dissolve their charms by our medicines. God has given us one little thing, which you know nothing of. He has given us the knowledge of certain medicines by which we can make rain. WE do not despise those things which you possess, though we are ignorant of them. We don't understand your book, yet we don't despise it. YOU ought not to despise our little knowledge, though you are ignorant of it.

MEDICAL DOCTOR: I don't despise what I am ignorant of; I only think you are mistaken in saying that you have medicines which can influence the rain at all.

RAIN DOCTOR: That's just the way people speak when they talk on a subject of which they have no knowledge. When we first opened our eyes, we found our forefathers making rain, and we follow in their footsteps. You, who send to Kuruman for corn, and irrigate your garden, may do without rain; WE cannot manage in that way. If we had no rain, the cattle would have no pasture, the cows give no milk, our children become lean and die, our wives run away to other tribes who do make rain and have corn, and the whole tribe become dispersed and lost; our fire would go out.

MEDICAL DOCTOR: I quite agree with you as to the value of the rain; but you cannot charm the clouds by medicines. You wait till you see the clouds come, then you use your medicines, and take the credit which belongs to God only.

RAIN DOCTOR: I use my medicines, and you employ yours; we are both doctors, and doctors are not deceivers. You give a patient medicine. Sometimes God is pleased to heal him by means of your medicine; sometimes not—he dies. When he is cured, you take the credit of what God does. I do the same. Sometimes God grants us rain, sometimes not. When he does, we take the credit of the charm. When a patient dies, you don't give up trust in your medicine, neither do I when rain fails. If you wish me to leave off my medicines, why continue your own?

MEDICAL DOCTOR: I give medicine to living creatures within my reach, and can see the effects, though no cure follows; you pretend to charm the clouds, which are so far above us that your medicines never reach them. The clouds usually lie in one direction, and your smoke goes in another. God alone can command the clouds. Only try and wait patiently; God will give us rain without your medicines.

RAIN DOCTOR: Mahala-ma-kapa-a-a! Well, I always thought white men were wise till this morning. Who ever thought of making trial of starvation? Is death pleasant, then?

MEDICAL DOCTOR: Could you make it rain on one spot and not on another?

RAIN DOCTOR: I wouldn't think of trying. I like to see the whole country green, and all the people glad; the women clapping their hands, and giving me their ornaments for thankfulness, and ululating for joy.

MEDICAL DOCTOR: I think you deceive both them and yourself.

RAIN DOCTOR: Well, then, there is a pair of us [meaning both are rogues].

The above is only a specimen of their way of reasoning, in which, when the language is well understood, they are perceived to be remarkably acute. These arguments are generally known, and I never succeeded in convincing a single individual of their fallacy, though I tried to do so in every way I could think of. Their faith in medicines as charms is unbounded. The general effect of argument is to produce the impression that you are not anxious for rain at all; and it is very undesirable to allow the idea to spread that you do not take a generous interest in their welfare. An angry opponent of rain-making in a tribe would be looked upon as were some Greek merchants in England during the Russian war. (Livingston 2006)

# 8 The Highest Pressures Ever Recorded—and Those Pesky Calculations

> Barometer, n.: An ingenious instrument which indicates what kind of weather we are having.
>
> Ambrose Bierce, *The Devil's Dictionary*

The public tends to think of most elements of weather as straightforward and simple, both to understand and measure. After all, rain, temperature, wind—the meaning of those weather features is clear to most people.

But the weather variable known as atmospheric pressure is a bit different—even when discussing the air pressures displayed on television's evening weather forecast.

Let's address the basics. First, air has mass; it takes up space. Gravity gives that mass of atmospheric gas a specific weight. We can measure the weight of that gaseous mass by the force that it exerts on its surroundings. Think of determining a person's weight by observing the amount that a seat cushion is crushed as they sit down. We call the measurement of atmospheric force 'air pressure.'

One important aspect of air pressure is that it changes with elevation. Lower elevations experience higher air pressures than do mountainous elevations. That difference is the reason that planes are pressurized as they travel. Commercial jets are flying at an altitude that experiences less than a third of the air pressure that most of us encounter on the ground. A similar situation occurs when a person is hiking in the mountains. There is less air pressure at the top of the mountain than at the bottom. A few years ago, when I stood at the top of Mount Kilimanjaro in Africa at an elevation of 5,895 meters (19,341 ft) above sea level, the air pressure was a bit more than half of what I had experienced standing at sea level.

Scientists like numbers. Can I express that atmospheric pressure at sea level or on the top of Kilimanjaro with specific numbers?

We measure air pressure using two different sets of units. The pressure scale with which most Americans are familiar is inches of mercury. That scale dates to one of the first instruments used to record air pressure.

In 1643, Italian physicist Evangelista Torricelli invented the first barometer. A simple device, it consisted of a tube resting in an open-faced dish filled with a liquid. The idea of the instrument was straightforward: air would push down

DOI: 10.4324/9781003367956-8

on the liquid in the dish, forcing some of the liquid into—and up—the tube. The distance that the weight of air pushes the liquid up through the tube is registered as a measure of air pressure.

Torricelli's first attempts involved a simple tube resting in water. How far would water be forced up the tube when the weight of air was pushing on it?

The problem is that water is relatively light in weight—and is rather easy for air to push against. So, a *very* tall tube with a large amount of water was needed. Given the density of water, air should push the water many feet up through the tube. Torricelli's experimental water barometer rose more than ten meters (33 feet) in height, standing well above the roof of his home! The sight of this odd device even caused some concern for Torricelli's neighbors, who wondered if the scientist was somehow involved with witchcraft.

To keep his experiments less voluminous (and less public), Torricelli realized that he could create a decidedly smaller barometer by using a much heavier—denser—liquid. The substance he chose for that new barometer was mercury. At normal temperatures, mercury is a silvery liquid that weighs fourteen times as much as a similar amount of water. So, the distance that mercury is pushed up a tube because of common air pressure could be measured in inches rather than feet.

At sea level, Torricelli's barometer would push mercury up the tube to a height that was a bit more than twenty-eight inches (71.1 cm). Thus, the inches of mercury pressure scale was created, a scale that is still used throughout the United States.

But, for meteorologists, the preferred measurement systems use units of pressure expressed in hectoPascals (hPa) or millibars (mb). At sea level, a barometer calibrated in those pressure units would read 1,013 hPa or 1,013 mb, respectively.

Air pressures can also change across horizontal distances. A massive hurricane in the Atlantic, for example, may have an air pressure under 900 hPa (26.58 inches of mercury), while a few thousand miles to the northeast an immense dome of cold polar air over Siberia might experience an air pressure well over 1,020 hPa (30.12 inches of mercury).

That causes a problem for meteorologists. If we want to create a regional or global map of air pressures so that we can forecast cold waves and hurricanes, we don't want to show the constant (nonweather) reductions in air pressures that are associated with elevation change. For example, we don't want to have a weather map consistently displaying a low barometric reading in mountainous Denver, Colorado compared to its sea-level counterpart in Key West Florida.

Meteorologists realized that we needed to standardize all air pressures to one single uniform elevation so that we are always comparing apples to apples. We want to only show the weather-related variations in pressure, like hurricanes and cold waves, and not pressure changes due to altitude.

Weather scientists selected the standard elevation for surface weather to be sea-level. And, to this day, the results of that decision are shown on the common pressure map displayed during the evening weather—a map showing all locations' pressures *as if everything was at sea level.* The *L*'s and *H*'s indicated on such a map are indicating regional low or high pressure, respectively, as if *at* sea level. But how can we convert a station air pressure measurement taken at high-elevation Denver, Colorado down to its sea level equivalent? After all, there is a mile of solid rock between Denver and sea level. How can we determine what Denver's sea-level pressure would be?

The answer involves some serious physics, mathematics . . . and averaging.

As part of any sea-level transformation formula, we must compute a sea-level pressure based on a determination of what the average change in temperature would be through a chunk of atmosphere, for example, the space from Denver's mile-high elevation down to sea level.

But how can we determine an average change in temperature between different elevations—what meteorologists call a 'lapse rate?'

In this case, we have taken decades' worth of data from the weather balloons launched every day over an area and averaged them together into something that we call the International Standard Atmosphere (ISA). The ISA is a single set of numbers representing the average conditions found in our atmosphere at various levels from the earth's surface upwards.

Continuing with our example of Denver, using that ISA information and Denver's actual measured surface pressure, we can then use a mathematical formula involving surface temperature and other values to compute what the hypothetical sea level pressure would be. The method performs equally well for a high-elevation place like Denver or a near-sea-level place like Miami.

As seen throughout more than a century of meteorological research, these mathematical calculations work most of the time. Every day around the entire world, the world's national weather services compute and create surface weather maps, using these adjustment equations, so that we can see the patterns in sea-level pressures around the world. Those air pressure patterns are fundamental to our knowledge—and forecasting—of weather.

But would those mathematical equations work for *extreme* atmospheric pressures?

On 30 December 2004—an *exceptionally* cold morning in the mountainous city of Tosontsengel in Mongolia—the meteorologists working in that remote place recorded a *station* pressure of 846.5 hPa. When they applied their mathematical adjustment formulae to standardize the observation down to sea level, they computed a sea-level pressure of 1089.4 hPa (32.17 inches of mercury) for the Tosontsengel reading.

Remember that our globally normal sea level pressure is 1,013 hPa. If we confirmed the Mongolian observation, that 1089 hPa value would be the highest sea-level pressure ever recorded for the planet. Incredibly high pressure!

Tosonstengel, Mongolia is situated in the northwest portion of Mongolia. The weather station is at an elevation of 1,724.6 meters above sea level (at 5,656 feet, roughly the same elevation as Denver, Colorado). The monitoring station has operated continuously since 1963. Geographically, Tosontsengel sits within a large valley surrounded by mountains—a valley into which cold air can sink, causing very low temperatures. On that brutal December morning in Tosontsengel, the thermometer read a bone-chilling value of -44.8°C (-48.6°F). The equipment used to record the barometric reading was a sophisticated mercury-based barometer.

My WMO evaluation committee consisted, as usual, of a very knowledgeable and well-qualified set of international scientists. We had respected experts from the United Kingdom, Spain, Argentina, and Morocco, as well as from the United States. Since the reading had been taken in Mongolia, it was appropriate to have a local representative from that country serve on our committee. In this case, it was my great pleasure to invite a young up-and-coming scientist, Dr. Gomboluudev Purevjav, onto our evaluation committee.

Gomboo, as he is nicknamed by his friends, has (now) long worked in Mongolia's Information and Research Institute of Meteorology, Hydrology and Environment as their Scientific Secretary. His primary research fields are focused on regional weather and climate modelling, atmosphere and biosphere interaction, climate variability change and extremes and assessment of climate change impact and vulnerability study. Since our evaluation work, he has been engaged in Mongolia's National Communication of Climate and has led their adaptation team.

Our Mongolian colleague was able to acquire a wealth of information for us. Photographs of the station, data from other nearby stations and even the actual log sheet of weather observations for that location (remember our investigation of the 1922 Libyan temperature discussed in chapters 3 and 4). Most importantly (recalling the pressure discussion above), Gomboo managed to get the precise mathematical equations that were used to adjust the Mongolian mile-high surface pressure down to its sea level equivalent.

To give you a sense of the mathematical complexity that we were dealing with, this is the specific pressure formula used in Mongolia to adjust their pressure observations to sea level:

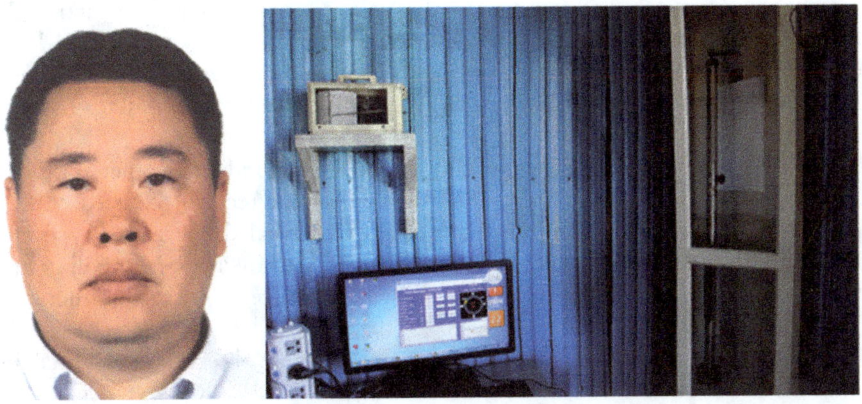

*Figure 8.1* Left, Dr. Gomboluudev Purevjav of the Mongolian Information and Research Institute of Meteorology, Hydrology and Environment. Right, photograph of the Tosontengel (Mongolia) meteorological station's SPA-A(B) mercury barometer.

$$P_o = P \exp\left(\left[10/\left\{R\left(T + 0.377\frac{e(T-273)}{P} + \frac{H}{2}\right.\right.\right.\right.$$

$$\left.\left.\left.\left.\left(\gamma + \alpha\left[0.377\frac{e(T-273)}{p}\right]\right)\right)\right\}\right]\phi \log e\right)$$

Where (to be fully scientific) $p$ is station pressure (hPa), $T$ is station air temperature (K), $H$ is the height of the station above sea level (m), $\phi$ is geopotential height of the station above sea level (gpm), $\alpha$ is a constant based on long-term temperature (°C), $\gamma$ is a constant (5.0), $e$ and $P$ are constants based on long-term pressure (hPa), and $R$ is the gas constant for dry air.

When we solved this equation for the local conditions at 8am local time on 30 December 2004, we computed a record sea-level pressure adjusted value of 1089.4 hPa.

But was it a legitimate observation?

First, as usual, we assessed the basics involving equipment and operating procedures. The Tosontsengel barometer had been a standard mercury-based instrument manufactured in Russia and the evaluation committee determined that the barometer had been working properly. Another pressure instrument at the location, a device known as a recording barograph, was even in general agreement with the primary instrument's barometric reading. The problem of exact replication is that a barograph determines *trends* of pressure within the last three hours and does not determine instantaneous pressure values. But its values were in line with the actual pressure observations.

We determined that the normal operational procedures had been followed. The Mongolian observer had taken proper manual measurements every three hours by visual reading using the scaling on the barometer. The barometer was located inside to prevent the mercury from freezing but the outdoor temperature was used in sea-level calculations—again, normal practice. We also established that the instrument had been properly maintained and checked. For example, every three years, the Tosontsengel station barometer had been compared with a calibrated reference laboratory barometer from the central office in Mongolia.

Finally, on that frigidly cold morning in 2004, surrounding weather stations in Mongolia also demonstrated very high (although not record) adjusted sea-level pressures. The weather maps created for the area revealed a large region of very high pressure in place over central Asia concentrated over the Mongolian area. Such a massive high-pressure system was consistent with the incredible Tosontsengel measurement.

Our standard concerns of verifying an extreme weather observation had been met.

But what about the conversion to sea-level pressure?

A problem became clear to the committee. There are more than *fifteen* different methods used around the world in reduction-to-sea-level formulae. Some have specific details to localized regions. For example, in the western United States and

Canada, the formulae contain a particular plateau adjustment to reduce the variability in the computed sea-level pressure. Of course, if the variability is reduced, the opportunities to achieve extreme sea level pressures are also reduced.

Was one method better than the others?

Even back in 1959, a theoretical meteorologist commented in one of the standard textbooks on meteorology that:

> [I]t should be clear that the methods for reduction to sea level are not uniform over the world and are especially complex in the United States. It would be desirable to make the procedure uniform and simple, but because many years of climatological records are based on the current unwieldy system it is unlikely that any revision will be made. . . All methods of reduction to sea level give unsatisfactory results in certain situations. (Hess 1959, 90)

Now, in the more than sixty years since then, that unwieldy system has, if anything, become more complex over time. And, yes, that old meteorologist's suggestion is still true that the selection of a single global reduction-to-sea-level adjustment equation for the world would remove some of our difficulties.

Wait a moment! Back in chapter 2, I said that the World Meteorological Organization had been created for precisely that reason—to standardize world observations. Isn't the WMO doing anything about it?

The short answer is yes . . . and no. At the time of writing, that issue is being addressed by a specific—different—body of the WMO, specifically the WMO Commission for Instruments and Methods of Observation (CIMO). But unfortunately (as if often the case with governmental organizations), they haven't yet reached a workable conclusion. As a start, the WMO has recommended a specific single-reduction formula to be used for *low-level* stations, those whose elevation is at or below 750 meters (2,460 feet). That's good, but Tosontsengel, Mongolia has an elevation of 1,724.6 meters (5,656 feet) above sea level, well above that 750-meter limit.

As a result, my committee was in limbo.

While we mulled over our options, I asked a colleague of mine, Dr. Bob Balling—a true statistical wizard of climatology—to take a close look at some of these sea-level reduction formulae. He found something a bit surprising.

When Balling examined the various variables in the reduction-to-sea-level equation, he found that, more than elevation, surface temperature was critical to determining computed sea-level pressure. Simple regression analysis (and other techniques) demonstrated that changes in surface temperature alone explain more than eighty percent of the variation in the computed sea-level pressures. That means statistics says that the bone-chilling temperature of -44.8°C (-48.6°F) was likely driving the very high computed sea-level pressure.

After a while, higher authorities in the WMO told me that it would likely be some time before there was any WMO recommendation of a single SLP formula for high elevations (and that statement has turned out to be correct—there still isn't any single recommendation now several years after this investigation). When we used the low-elevation WMO equation mentioned above, we arrive at a value of 1089.1 hPa (compared to the Russian formula's 1089.4 hPa). Therefore, I laid out three options to my committee:

a Although instrumentation and data collection procedures were properly followed, we could reject the Tosontsengel, Mongolia's extreme SLP of 1089.1 hPa as a world record based on the oddity of the station's localized observation (e.g., extreme low temperature and/or extreme high elevation).
b Although instrumentation and data collection procedures were properly followed, we could reject the Tosontengel, Mongolia's extreme SLP of 1089.1 hPa as a world record on the basis that current WMO policy guidelines do not specify which SLP reduction formula to use above 750 meters.
c Or we could accept the Tosontsengel, Mongolia's extreme SLP of 1089.1 hPa as a world record, but distinguish it from other SLP extreme observations that do meet current WMO policy guidelines and explicitly state potential caveats associated with its acceptance.

After discussion, we rejected options (a) and (b) as we determined that such actions could 'bias, infringe or hinder the on-going revision of existing WMO guidance' (yes, by this time, I was learning administrative jargon). In basic terms, if the Tosontsengel, Mongolia's extreme SLP was rejected on the basis of unrepresentative temperature or elevation, then we would have to define exactly what temperature/elevations limits were. And, that would involve setting policy for which another WMO committee held jurisdiction and which was (and still is) evaluating.

It was the unanimous recommendation of my WMO evaluation committee that the WMO should accept the Tosontsengel, Mongolia SLP pressure observation. They considered the observation to be a properly conducted observation that can be accepted as a world extreme SLP but with several caveats.

One of the major qualifications was that for weather stations above 750 meters, any record must explicitly mention any abnormally cold air temperature and that point should be explicitly addressed in extremes identification and verification.

Second, the WMO World Archive of Weather and Climate Extremes created two distinct sea-level pressure categories for observation of extreme measurements, one record for stations above and another record for below 750 meters. This, we believed, would help discriminate any pressure records for scientists as well as the public.

The end results? The WMO Archive for Weather and Climate Extremes now lists (a) Highest Adjusted Sea-Level Pressure (Below 750 meters) with an official observation of 1,083.3hPa recorded on 31 December 1968 at Agata, Evenhiyskiy, Russia (at an elevation of 261 m, 856 ft) and (b) Highest Adjusted Sea-Level Pressure (Above 750 meters) with an official observation of 1,089.1 hPa on 30 December 2004 in Tosontsengel, Mongolia.

As always, we added the final advisory note: in the future, if a single reduction-to-sea-level formula is globally accepted or if the WMO advisory committee makes another recommendation, we at the WMO World Archive for Weather and Climate Extremes may reevaluate this and other record extremes through another evaluation of international experts.

Ah, the pressures of working with atmospheric pressure!

# Interlude: Freaks of Pressure

This chapter has highlighted some of the problems with complex pressure measurements. It is only fitting that this interlude reviews a case of unconventional, out-of-box thinking involving academic pressure and the use of barometers.

This droll story was told by Dr. Alexander Calandra, a physical sciences professor at Washington University in St. Louis, Missouri. As he related in a 1964 issue of *Current Science*, the physics professor was once asked to judge a disagreement between a student and another professor regarding the grading of an examination question.

The colleague stated that the student should be given a failing grade of a zero for his answer to a physics question, while the student claimed he should receive a perfect score.

The examination question was 'Show how it is possible to determine the height of a tall building with the aid of a barometer.'

The expected, conventional answer was that atmospheric pressure at the bottom and top of the building would be different (higher pressure at the bottom, lower pressure at the top) and that difference could be measured by the barometer. With precise measurement of the pressure difference and knowledge of the mathematical relationship linking height and pressure, a person could calculate the building's height.

In contrast to that expected answer, the student's response was:

> Take a barometer to the top of the building, attach a long rope to it, lower the barometer to the street and then bring it up, measuring the length of the rope. The length of the rope is the height of the building. (Calandra 1964)

Calandra pointed out to his colleague that the student did have a good case for receiving full credit since he had answered the question completely and correctly. On the other hand, Calandra then told the student, a high grade was supposed to demonstrate that the student knew and understood some physics, but his written answer to the question did not confirm such knowledge.

So, as a referee to the situation, Calandra proposed that the student be given another try at answering the question. To the physics professor's surprise, both the colleague and the student agreed to the retest.

As mentor to this second test, Calandra gave the student six minutes to give another answer to the barometer question—with the stipulation that the answer this time must show some knowledge of physics. Five minutes passed and the student hadn't written anything on his paper.

Concerned, Calandra asked if he wished to give up, but the student replied no, he wasn't giving up. He explained that there were many different answers to this particular problem, and he was choosing which one to use. In the next minute, he dashed off his answer.

Calandra read:

> Take the barometer to the top of the building and lean over the edge of the roof. Drop that barometer, timing its fall with a stopwatch. Then, using the formula $S = ½\ a\ t^2$, calculate the height of the building. (Calandra 1964)

At this point Calandra probably smiled—at least, I would have—and found in favor of the student.

As the physics professor and the student left the colleague's office, Calandra remembered that the student had considered many other solutions. The professor asked the student what other solutions were possible.

'There are many ways of getting the height of a tall building with the aid of a barometer,' the student had responded:

> For example, you could take the barometer out on a sunny day and measure the height of the barometer, the length of its shadow, and the length of the shadow of the building, and by the use of simple proportion, determine the height of the building.
>
> Or there is a very basic measurement method that you will like. In this method, you take the barometer and begin to walk up the stairs. As you climb the stairs, you mark off the length of the barometer along the wall. You then count the number of marks, and this will give you the height of the building in 'barometer units.' A very direct method.
>
> Of course, if you want a more sophisticated method, you can tie the barometer to the end of a string, swing it as a pendulum, and determine the value of g at the street level and at the top of the building. From the difference between the two values of g, the height of the building can, in principle, be calculated. (Calandra 1964)

The student finished:

> If you don't limit me to physics solutions to this problem, there are many other ways. Probably the best is to take the barometer to the basement and

> knock on the superintendent's door. When the superintendent answers, you speak to him as follows: 'Mr. Superintendent, here I have a fine barometer. If you tell me the height of this building, I will give you this barometer.' (Calandra 1964)

In science, I tell my students, unconventional thinking is something we should strive for. Always think outside of the box. A bit later on, when I discuss lightning extremes in chapter 12, I tell you of an incredible researcher who epitomizes this type of innovative thinking.

Finally, with regard to the freaks of atmospheric pressure, I must mention a favorite movie of mine, the 1995 Bruce Willis action adventure *Die Hard with a Vengeance*. In the middle of the film, the hero John McClane has followed the bad guys underground and into an ongoing construction project. That project involved the creation of a new aqueduct for New York City. After defeating several of the enemy, McClane finds that some of the aqueduct tunnel partitions—critical ones holding back the turbulent river waters—had been removed in a dastardly attempt to kill him. Water is surging through the massive passageway. In the nick of time (of course), McClane manages to jump up to a ladder connected to a sewer tunnel, which leads up to the surface. However, before our hero is completely out of the shaft, the pressure of the surging waters reaches him and miraculously McClane is catapulted high into the air and tumbles down to the ground, next to a freeway—unharmed and ready for the next thrilling action sequence.

Only an improbable action movie stunt? And how does atmospheric pressure become involved?

In 1916, a 'sand hog'—the term used for a tunnel worker during the construction of the New York Subway—named Marshall Mabey was drilling in a deep tunnel under the East River on the Whitehall-Montague Street spur of the Dual System subway. That section of the subway had been heavily pressurized with air to help hold back water leaking in from the river above. Even so, a weak section holding back the river bottom would occasionally give way under that heightened air pressure, causing the subway tunnel to experience a catastrophic 'blowout.'

As Mabey was working on his section of the new shaft under the East River, the compressed pocket of air about him exploded up through the bottom of the river like an erupting cork of air. Mabey and the other two sand hogs were sucked up in a cataclysmic blowout of air, water and dirt. Witnesses on the shore observed an immense geyser of air and water, estimated to be fifty feet (fifteen meters) high, bursting as if by magic from the surface of the river.

On the top of that fountain of water and air was the dancing figure of a man, tumbling in the turbulent spray. It was Mabey. Miraculously, the sand hog was rescued and found to be almost unscathed by his explosive experience. Unfortunately, the other two men with him weren't so lucky. Both perished in the blowout from severe head contusions.

Yes, truth can be as strange as movie fiction!

# 9 The 'Largest' Hail but No Snow

> The thin snow now driving from the north and lodging on my coat consists of those beautiful star crystals. . . How full of the creative genius is the air in which these are generated! I should hardly admire more if real stars fell and lodged on my coat.
>
> Henry David Thoreau, *The Journal of Henry David Thoreau, 1837–1861*

Snow and hail have one thing in common: their origins generally both reside in the odd properties of ice crystals.

Ice crystals are the 'seeds' of most snowflakes and hailstones. Such crystals display a six-sided symmetry because on a large scale they replicate the unvarying structure of the crystal's water molecules—two hydrogen atoms linked to an oxygen atom. The crystals arrange themselves in predetermined spaces (a process known as 'crystallization') to form a six-sided snowflake.

The exact type of ice crystal that forms is the result of the specific temperature and the humidity of the atmosphere. For example, long needle-like crystals form when the air temperatures are at 23°F (5°C), while very flat plate-like crystals form around 5°F (-15°C). No matter the temperature, all ice crystals keep the basic hexagonal shape. The crystal maintains that shape as it falls—and grows—through the atmosphere.

The expanding shape of an ice crystal is a product of the ever-changing atmospheric conditions that the crystal encounters as it falls through tens of thousands of feet of air. For example, a crystal might begin to develop arms off its six sides in one manner, and then minutes or even seconds later, minor changes in the surrounding temperature or humidity causes the crystal to grow in another way. Although the hexagonal shape is always maintained, each of the ice crystal's six arms may branch off in new directions. Because a growing crystal's arms experiences the same atmospheric conditions, the arms of the crystal itself appear identical—but the slight variations in the atmospheric conditions existing between the ensuing snowflake and its neighbors ensures it will be different from its neighbors.

Many people have speculated on this intriguing uniqueness inherent in snowflakes. As with many aspects of weather, the subject of snowflakes points out the surprising number of superstar past scientists who have worked in meteorology. For instance, the great German scientist, Johannes Kepler—the

DOI: 10.4324/9781003367956-9

first person to determine that planets travel about the sun in elliptical orbits, also was one of the first people to examine snow crystals with a scientific eye. In 1611, Kepler wrote an intriguing but brief paper (as a New Year's gift for a friend) entitled 'The Six-Cornered Snowflake,' in which he addressed the individual character of snow crystals in relation to a reasoning process. In comparing flowers and snowflakes, Kepler wrote:

> Each single plant has a single animating principle of its own, since each instance of a plant exists separately, and there is no cause to wonder that each should be equipped with its own peculiar shape. But to imagine an individual soul for each and any starlet of snow is utterly absurd, and therefore the shapes of snowflakes are by no means to be deduced from the operation of soul in the same way as with plants. (Libbrecht 2007, 54)

Despite the odd language, Kepler was essentially correct in his thinking. There is no *biological* blueprint for snow-crystal development. They are 'soulless.' Their growth is determined by consistent physical rules—which are far simpler than the genetic chemistry of living organisms—yet complex shapes emerge spontaneously. Of particular interest to me, it appears that Kepler might have been channeling future thoughts of modern fractal theory as he pondered the genesis of complex patterns and structures from simple precursors. It is a worthy scientific question, and one that scientists are still investigating today.

Ice crystals can develop and grow because they exist in a vast aerial lake—a cloud—with other ice crystals and with countless microscopic water droplets. Because of their small size, these trillions of infinitesimal water droplets can remain liquid in what is called a 'supercooled' state. Surprisingly, supercooled water can stay liquified to temperatures as low as forty degrees below zero (uniquely, both in Fahrenheit and Celsius). It is these supercooled water droplets that provide the moisture for ice to grow on the suspended ice crystals. Roughly one million microscopic, supercooled cloud droplets are needed to provide enough water vapor to grow a single large snow crystal.

The snow crystals become heavier as they grow, until gravity pulls them out of their cloud-like nurseries. As they fall, if the temperatures warm to above freezing, the ice crystals melt into raindrops. But, if the air temperatures stay below freezing, the ice crystals continue to fall and form snowflakes. Each one is the unique result of its intricate pathway down through the thousands of feet of that aerial cloud lake.

Through science, we understand snow. But one aspect of the atmospheric sciences, which I also love, is that weather can also be appreciated for its utter beauty.

People have long been captivated by the wonder of snowflakes and some have thought to preserve that splendor. For over forty years in the late 1800s and early 1900s, the renowned Wilson 'Snowflake' Bentley photographed thousands of individual snowflakes and, as a result, he perfected innovative photomicrographic techniques that are still used for documenting minute objects. His multitude of snowflake photographs have provided valuable scientific records of snow crystals and their many types. Indeed, the Smithsonian Institution Archives still contains five hundred of his unparalleled snowflake photos—images that Bentley offered to the Smithsonian in 1903 to protect against 'all possibility of loss and destruction, through fire or accident' (Anonymous 2023).

Conclusion: without question, snow is fascinating, is beautiful and is of scientific interest.

So why don't we list snowfall extreme records in the WMO World Archive of Weather and Climate Extremes?

Ouch. I get asked that question a lot.

The answer has to do with that ever-changing nature of snowflakes.

An essential part of any snowflake is—surprisingly—air. Upwards of ninety percent of a snowflake's total volume can consist of air. And that amount of air within a snowflake becomes changeable once the snowflake falls to earth. When a fallen snowflake lies on the ground, new snowflakes will fall onto it, and newer snowflakes will fall onto them.

As the weight on the original snowflake increases, that flake becomes compressed—and some of that air between, and within, the snowflake's arms squeeze out of it. The snow becomes denser.

As a side note, that compression is a continual process. For example, snow that falls on a glacier is compressed by new snow falling on it, which in turn is compacted by newer snow and so on, until, within a period of twenty years or so, the fallen snow is transformed into a much different, *denser* material called glacial ice. When many years ago, I was in Antarctica extracting an ice core, I discovered that such compressed glacial ice can have the consistency of concrete—a sledgehammer won't even break it!

But even in the short term, a snowflake's loss of air through compression is important in the calculation of snowfall. In part, the reliability of snowfall measurements is a product of how frequently those measurements are made.

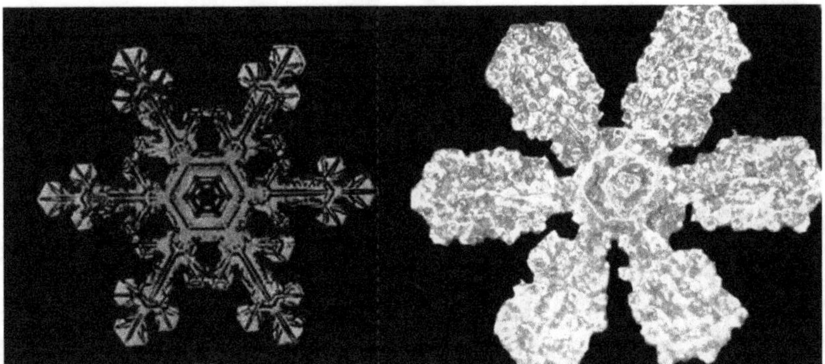


*Figure 9.1* The beauty of snowflakes. Left, a classic Wilson Bentley 1901 photograph of a pristine snowflake; right, a photograph of a 2023 rapidly melting Antarctic snowflake by the author.

As mentioned in chapter 2, this point was critical in the evaluation of a potential record measurement of snow taken during a lake-effect snowstorm near Buffalo, New York in January of 1997. In that storm, a snow spotter reported six measurements of snow over the course of twenty-four hours, which when summed totaled an amazing seventy-seven inches (1.96 meters).

Officials from the US National Oceanic and Atmospheric Administration examined the observation (in fact, it was this investigation that spurred the formation of the US National Climate Extremes Committee). First, the expert meteorologists determined that the observer took the measurements in an excellent location for snow measurement: open from all directions, not affected by any buildings and with enough nearby trees to result in little drifting of snow.

Unfortunately, there isn't any acceptable automated device to measure snowfall. In the New York state situation, manual measurements use a standard, two-foot-wide snow board. Typically, a snow board is composed of a flat piece of thin metal, wood, or other material, light so it won't sink in the snow, and painted a white color. The board helps to reduce error. Measurements of snowfall, if measured on grass, can be exaggerated as the grass blades themselves can create inflated snow totals. The white color of a snowboard serves to reduce heating by sunlight, which often occurs on paved surfaces.

The observer in western New York had a good snow board.

Next, the observer needed to take a measurement. Snowfall measurement would seem to be straightforward. Push a calibrated ruler (such as a yardstick or meter stick) straight into the snow, perpendicular to the ground, until the yardstick reaches the snow board. Then record the measurement (in the United States) to the nearest tenth of an inch; for example, one might measure 3.3 inches. Keep track of all measurements for the duration of the storm to compute the storm total amount. Finally, after recording a measurement, clear off the snow board (then place it on top of the snow), so it is ready for more snow.

When the NOAA officials evaluated this snowfall extreme, they found that the observer in western New York followed those guidelines.

But here we had a problem. The observer in New York State conscientiously recorded six measurements between 1:30 pm on 11 January and 1:30 pm on 12 January 1997, a twenty-four-hour period. But he made five of those measurements within a single twelve-hour period.

The National Weather Service notes that an observer should 'not measure every hour and add them up . . . this would give an unrealistic high amount [for climatological purposes], [one reading] every 6–12 hours would be fine (unless it is melting)' (National Oceanic and Atmospheric Administration 1997, ES-1).

Why would they stipulate that?

We must allow gravity and future snow to compress naturally the snow. Some of the air must have a chance to be expelled from the compressing snowflakes. The WMO's guidelines are even more restrictive, advising that snow:

> . . . be allowed to collect undisturbed on the snowboard during the observation period. Observations are to be made at the same time each day [thereby

> implying one snowfall measurement per day]. The observer should measure snowfall on the board using a ruler to the nearest 0.5 cm. (World Meterological Organization 2023, 10)

The conclusion of the committee? The observer in western New York made too frequent of measurements and therefore they shouldn't be combined into a single value. As the NWS noted, 'More frequent measurements . . . tend to increase totals, especially when fluffy snow [such as that lake-effect snowstorm] are involved.'

That observer was a respected NWS snow spotter for the Buffalo, New York National Weather Service Forecast Office. In fact, the official review said he was 'extraordinary diligent.' But his too-frequent records were then inappropriately combined to claim a record snowfall event—a record that was broadcast across global media, including television's Weather Channel. Consequently, if records of a trained spotter in a country such as the United States can be misinterpreted, unfortunately, the likelihood of snowfall error across the world remains high.

So, I have made the decision *currently* not to include snowfall records as part of the WMO Archive of Weather and Climate Extremes.

That decision was brought up in 2015, when officials in Italy announced a record twenty-four-hour snowfall of 100.8 inches (256 centimeters) in the small mountain town of Capracotta, located three hours southeast of Rome in the Italian Apennines. Most of the world media added a cautionary statement about snowfall measurements. For example, the Washington *Post* noted,

> An investigation of the measurement by the World Meteorological Organization would need to be conducted in order for this to go down in the 'official' record books, but the WMO does not currently track snowfall for any location. According to Randall Cerveny, WMO's chief rapporteur of weather and climate extremes, this is because accurate snowfall measurements are fairly limited and have been 'markedly difficult' to verify. (Fritz 2015)

And that difficulty continues even today. Presently, the WMO Archive of Weather and Climate Extremes does not include snowfall records. Misquoting J. R. R. Tolkien from the *Lord of the Rings*: 'A day may come when the WMO will accept snow records from around the world, but it is not this day.' It is the current position of the WMO Archive of Weather & Climate Extremes that the potential problems in verifying extreme snowfall measurements outweigh the worth of the observations.

A similar situation exists with hail.

Hail—at its most basic—is an ice crystal on steroids. And in this case, the natural steroids for creation of a hailstone are updrafts.

In my earlier discussion of snow, I mentioned that the snow crystal's become heavier as they grow, until gravity pulls them out of their cloud-lake nurseries.

What happens if something happens to prevent gravity from pulling them down?

The key to creating hail is updrafts. Strong updrafts can make big hailstones.

One of the basic rules of meteorology is that hot air rises. Hot air is less dense than cold air and, therefore, heated air rises. As it ascends, that air cools and the moisture in it will condense into ice crystals and water droplets. The micrometer-sized oblong spheres of ice and water are effective at scattering light, with vast numbers allowing for maximum scattering, thereby creating visible clouds. A good-sized cloud bank might contain a million or so *tons* of water—a literal atmospheric lake, all in the form of suspended water droplets.

Hailstones form when those water droplets are carried upward by the warm thunderstorm updrafts into the higher, colder areas of the atmosphere and freeze. Hailstones grow by colliding with supercooled microscopic liquid water droplets whose moisture freezes onto the hailstone's surface.

The longer storm updrafts hold up a hailstone, the larger it can grow. The conditions—and the growth—experienced by the hailstone changes as it is held aloft by an updraft. Hailstones can have alternating layers of clear and cloudy ice as they pass through different temperature and liquid water content conditions in the thunderstorm. Imagine the hailstone is on the greatest roller coaster or bumper car ride of all time—one that is not only moving the hailstone up and down but even back-and forth through different temperature and humidity zones.

During this wild roller coaster ride, if the hailstone collides with the supercooled water and that water freezes instantaneously onto it, cloudy ice will form. This is due to air bubbles being trapped within the newly formed ice. But if the supercooled water freezes a tad more slowly, the air bubbles can escape and the new ice on the hailstone will be clear. A person can even witness these differences. If you cut a hailstone in half, you'll see these distinct layers—clear and cloudy—showing the hailstone's path through the storm's environment. These hailstone layers can resemble growth rings like those of a tree.

Ultimately, the hailstone plummets to the ground when the thunderstorm's updraft can no longer support the growing weight of the hailstone. That can occur if the icy stone becomes large enough or the updraft weakens. The hailstone's fall speed depends on a variety of factors, such as its size, the friction between it and surrounding air, the local wind conditions, and—a factor that will become important below—the degree of melting of the hailstone.

Early research assumed that hailstones fell like solid ice spheres. Scientists computed very high fall speeds, even for very small hailstones. More recently, studies using 3-D printed casts of real hailstones suspended in a vertical wind tunnel has demonstrated that natural hailstones fall more slowly than solid ice spheres. For small hailstones, less than an inch (2.5 cm) in diameter, the expected fall speed is between nine and twenty-five mph (4–11.2 m/s). For hailstones that one would typically see in a severe thunderstorm (one inch to 1.75 inch in diameter, 2.5 to 4.4 cm), the expected fall speed is between twenty-five and forty mph (11.2–17.9 m/s). In the strongest supercells that produce some of the largest hail one might expect to see (two inches to four inches in diameter, 5.1 cm to 10.2 cm), the expected fall speed is between forty-four and seventy-two mph (19.7–32.2 m/s). There is much uncertainty in these estimates due to variability in the hailstone's shape, degree of melting, fall orientation, and the environmental conditions. It is even possible for very large hailstones (diameters exceeding four inches, 10.2 cm) to fall at over one hundred mph (44.7 m/s)—as *the* authority for hailstone research (Dr. Charles Knight, whom I discuss below) demonstrated in 2019, when he tested hailstone fall rates by actually dropping hailstone mockups from aircraft.

One consequence of that speed is that hailstones can kill people.

In chapter 13, I'll discuss the world's worst hailstorm death toll, which happened in 1888, but even in today's modern world, hailstones still can kill. As recently as late August of 2022, a large hailstone killed a child only twenty-months-old during a severe thunderstorm that struck in northeastern Spain. More than fifty other people in the area suffered injuries—the majority from broken bones and bruising.

Hail doesn't occur uniformly around the world. For example, in the United States, the states on the lee side of the Rockies—Nebraska, South Dakota, Colorado and Wyoming—tend to have the most hailstorms, with as many as seven to nine hail days per year. Other parts of the world that suffer from damaging hailstorms include China, Russia, India, Spain and northern Italy.

And, of course, that means one question is often asked of me.

How big can hail get?

That depends on how you measure it.

The two common ways of measuring hail are by its size (diameter and/or circumference) and by its weight. In popular media, the first tends to be the one most often reported. But my experts on hail have recommended to me that the WMO only use the second—a hailstone's weight—in charting an extreme.

Why?

Because a hailstone's size changes too quickly after falling.

Hail consists of frozen water—which melts. Even before it hits the ground, a hailstone is melting. Most measurements of hail's size are conducted *after* the hailstorm is over by comparison against a known object. For example, the US National Severe Storm Laboratory lists these examples of comparative measures for US measurement:

Pea = 1/4-inch (0.6 cm) diameter
Mothball = 1/2-inch (1.3 cm) diameter
Penny = 3/4-inch (1.9 cm) diameter
Nickel = 7/8-inch (2.2 cm) diameter
Quarter = 1 inch (2.5 cm) — any hail that is quarter-size or larger is labeled 'severe'
Ping-Pong Ball = 1 1/2-inch (3.8 cm) diameter
Golf Ball = 1 3/4 inches (4.4 cm) diameter
Tennis Ball = 2 1/2 inches (6.4 cm) diameter
Baseball = 2 3/4 inches (7 cm) diameter

Tea cup = 3 inches (7.6 cm) diameter
Softball = 4 inches (10.2 cm) diameter
Grapefruit = 4 1/2 inches (11.4 cm) diameter

Of course, direct measurements of a hailstone's size using a ruler, calipers or a tape measure are best. But even with good measuring equipment, that pesky melting problem remains. The shape of the stone will be constantly changing—and even the use of one's hands in handling and measuring the ice stone will enhance that melting. The hailstone's measurements will *always* be changing.

Dr. Charles Knight urged me to evaluate hailstone extremes based only on their weight. Hailstone weight will not change as quickly, or as much, as the hailstone's size. To me, Knight's opinion on hail is equivalent to a decree handed down from the heavens. As a scientist at the US National Center for Atmospheric Research (NCAR) since 1962, Knight is *the* world-recognized authority on hail and supercooled water.

Quite simply, the illustrious Charlie Knight—along with his late wife Nancy—has long been regarded as one of the most knowledgeable experts in hail research in the world, with years of experience in the laboratory and in the field. I had the great honor of working with both of the Knights many years ago when we three wrote an article for the popular weather magazine *Weatherwise.*

Given his valued opinion, we at the WMO evaluate hail only by its weight, not size.

In 2010, we evaluated a candidate for that hail weight category. At that time, the WMO investigated a report of a hailstone recovered from the small town of Vivian, South Dakota, on 23 June 2010. Measurements revealed that the stone weighed 1.94 pounds (1 pound, 15 ounces) or 0.879 kg. If verified by the WMO ad-hoc evaluation committee, this event would rank as the heaviest hailstone ever recorded for the Western Hemisphere.

For this investigation's committee, I was fortunate to have that best of the best in hailstone research—Dr. Charles Knight—as a member of the evaluation team. Additionally, I had a good friend, Dr. Deke Arndt, who is now Director of the NOAA National Centers for Environmental Information, as well as other illustrious members from India, Spain, Morocco, and Argentina.

We were also fortunate in this case to have detailed information on the occurrence already collected by the US National Climate Extremes Committee, then under the direction of Deke Arndt.

The hailstone fell from a supercell storm—a thunderstorm that is so severe that the storm itself is rotating. There was abundant warm, moist air at the surface and radar indicated strong wind shear (winds coming in from different directions)—both factors that promote uplift, a key ingredient for hail in the storm.

That this storm produced severe hail is without doubt. Officials reported five hail-related injuries in the wake of this storm. According to local authorities and the regional meteorologists, the town of Vivian experienced extensive structure damage including broken windows and holes in roofs due to the large hail.

Also, according to local officials, the largest stone measured eight inches (0.203 m) in diameter with an 18 5/8 inch (0.473 m) circumference. The local weather representative confirmed that, upon measurement, it appeared that some more melting had taken place. The observer did keep the stone in a sealed plastic bag and kept the hailstone refrigerated. Several days later, the stone was measured on a scale provided by the US Post Office in Vivian. The measured weight was 1 pound, 15 ounces (0.879 kg).

As the report concluded:

> It is almost certain that this stone was larger upon impact than it was at the time of measurement and documentation. The combination of outdoors exposure before retrieval, repeated handling, and power loss certainly contributed to melting or sublimation of the stone. (Enloe 2010, 1)

By having the stone *inside* a sealed plastic bag, the total mass—the frozen stone and any melted bits—should be consistent to the time of the procurement and initial measurement. Therefore, my WMO committee recommended that the 2010 Vivian, South Dakota hailstone be accepted as the heaviest recorded hailstone for the Western Hemisphere, with a weight of 1.94 pounds (1 pound, 15 ounces) or 0.879 kg. I accepted that recommendation.

One might wonder with a nearly two-pound hailstone as the Western Hemisphere's heaviest recorded hailstone, what is the *world's* heaviest hailstone ever recorded?

That record belongs to a hailstone that fell in 1986 in Bangladesh. That stone weighed an incredible 1.02kg (2.25lb). It shouldn't be surprising to learn that the hailstorm that produced such a massive icy stone was also a killer. Ninety-two people died in the area because of that deadly hailstorm.

Hail can be amazing . . . and deadly.

*Figure 9.2* The 2010 Vivian SD hailstone, the 'heaviest recorded hailstone' for the Western Hemisphere, with a weight of 1.94 pounds (1 pound, 15 ounces) or 0.879 kg.

Source: Photograph courtesy of Aberdeen South Dakota National Weather Service Weather Forecast Office.

# Interlude: Freaks of Snow and Hail

I mentioned in this chapter that I had the distinct honor of working with Drs. Charlie and Nancy Knight when we three wrote an article for the popular weather magazine *Weatherwise*. One topic that particularly fascinated us was the idea of *inclusions* in hail—those strange 'seeds' that can be the nuclei of hailstones.

Sometimes strange impurities can seep into hailstones *after* they fall. For example, the Knights once discovered an Oklahoma hailstone that contained bright red-orange dirt. Microscopic analysis indicated that a tiny channel or opening in the hailstone likely allowed muddy water to enter the stone *after* it had fallen to the ground. Such tiny channels may account for stories of red or blood hailstones, such as fell in Italy in 1873 or in Russia in 1880.

To produce some hailstones winds can lift small objects high into the atmosphere. Those tiny objects can become the condensation nuclei—the natural seeds—of a hailstone. Some of those objects can be surprising. As a classic example, one of my favorite photographs (published in my book, *Freaks of the Storm*) was sent to me by Theresa DeBoer, showing an assortment of Nebraska hailstones, one of which had an actual ladybug embedded in it!

Hail can also be economically devastating. The pounding force of hailstones can destroy an entire year's crop within minutes. In one strange case, scientists have taken the idea of hail's condensation nuclei, and tied it to one valuable crop grown in Africa. They suggest that this important crop might supply the very seeds of its own destruction.

The Kericho area of Western Kenya is prized for the production of tea. It is also known for its disastrous hailstorms, with an astounding average of 132 days of hail occurring every year. In 1982, the husband-and-wife research team of R. C. Schnell and Suan Tan-Schnell discovered that the explanation for so many days of hail in Kenya might be the very production of the tea itself.

They observed that the organic ground litter from the tea-growing plants, if lifted into the sky by strong winds, could provide an abundant source of condensation nuclei and create hailstones! The Schnells' work in Kenya revealed that many of the icy stones from hailstorms had, at the core of the hailstones, tiny fragments of tea leaves. In essence, the Schnells discovered that the tea plantations in Kenya were supplying the very ammunition needed for their own destruction by hail!

Other hailstone seeds can be much larger—and deadlier.

During a gliding competition taking place before the start of World War II, five glider pilots elected to bail out when their gliders were caught in a severe thunderstorm over the Rhön Mountains in Germany. When the pilots abandoned their gliders, the thunderstorm's updrafts yanked the pilots' parachutes skyward. As they were carried up and down through the storm, the glider pilots were encased in ice, much like hailstones.

The consequences were dreadful. All five were covered with layers of ice. When they finally fell out of the storm and crashed to the ground, only one of the glider pilots survived and he lost three of his fingers and much of his face, due to the freezing ice. Tragically, the others had become frozen corpses plummeting to the ground.

But to end on a somewhat happier note, there is an occurrence in history where a massive hailstorm actually *ended* a war.

In the spring of 1360, during the Hundred Years' War, English King Edward III's army was in France, near Chartres, France. Their French opponents, low in numbers and led by the Abbot of Cluny, Androuin de La Roche, took shelter behind fortifications, and a siege ensued.

The next day, on Easter Monday of 1360, the English army made camp outside Chartres, in an open plain, when a violent thunderstorm plummeted the English with hail the size of 'goose eggs' (Froissart 1961, 72).

According to the *Old Chronicle*, the 'hailstones [were] so prodigious as to instantly kill 6,000 of his horses and 1,000 of his best troops' (Froissart 1961, 72). Even the armor that some English wore didn't offer enough protection from the icy attack—many dying from freezing exposure to the cold (a surprisingly common attribute of hail death). The hailstones had battered the troops into unconsciousness. The solders fell onto the icy-cold ground and froze to death. Hundreds of the English fighters were directly slain by the icy missiles.

Seeing the devastation and realizing that someone was trying to tell him something, King Edward III sprang from the saddle, stretched arms towards the Cathedral of Chartres, and vowed to God and the Virgin Mary that he would no longer object to honorable proposals of peace.

The next day, de La Roche arrived at the English camp with a set of peace proposals. Edward agreed to the terms and immediately withdrew his army from the gates of Chartres, concluding the one-day siege of the town.

A hailstorm had ended the battle.

# 10 Tropical Cyclones, the Planet's Strongest Storms

> The only embarrassing episode would have been engine failure, which, with the strong ground winds, would probably have prevented a landing, and certainly would have made descent via parachute highly inconvenient.
>
> The first 'hurricane hunter' Colonel Joseph P. Duckworth, after flying his B-25 through the eyewall of a Gulf hurricane in 1943

Tropical cyclones are the most powerful storms that occur on earth. These circular ocean storms are driven by immense releases of rising heat, through the conversion of water vapor into rain. Tropical cyclones are natural gigantic heat pumps that actually help to redistribute energy around the planet.

They can also be very, very deadly.

Our *modern* interaction with these powerful ocean storms dates to as early as 1274 and 1281 CE. According to stories from that time, in each of those years, Kublai Khan, one of famous rulers of the Yuan (Mongol) dynasty, launched two titanic armadas of warships from China, in separate attempts to conquer the island nation of Japan. In both cases, Japanese legends state that Khan's massive fleets were utterly destroyed by a '*Kamikaze*' (divine wind) typhoon. According to the Japanese legends, the two tropical cyclones were heaven-sent to protect Japan from invasion. Are those stories mere fantasy?

Recently, researchers have uncovered new physical evidence of extreme coastal flooding in Japan, resulting from the passage of massive typhoons that date to that time. Geologic sediment analyses from a Japanese coastal lake near the location of the legendary invasion site has provided direct evidence of the reality of those two tropical cyclones. The collected sediment has proven to be consistent with the stories of the two *Kamikaze* typhoons being of fearsome intensity. The evidence actually supports the historical accounts that the two storms played a critical role in preventing the sea invasion of Japan by the Yuan Dynasty fleets.

Hurricanes—typhoons' tropical counterparts in the Atlantic Ocean—have also long wrought death and destruction throughout that region of the world. Dave Ludlum, one of meteorology's premier weather historians, first described how Christopher Columbus's small fleet encountered a tropical cyclone in 1495 during his second voyage to the Americas. Available documentation doesn't

DOI: 10.4324/9781003367956-10

allow us to determine if the famous admiral was hit directly by a tropical storm or if he only experienced the outer fringes of a stronger hurricane near the island of Hispaniola. But it is recorded that the sea-tested admiral had correctly identified the first signs of the advancing storm and took precautions to safeguard his fleet. He anchored his ships safely behind Saona Island, off the southwestern tip of Hispaniola.

The knowledge that Columbus gained in that skirting encounter with a tropical cyclone served him well seven years later in 1502. When the then-aging admiral arrived at Hispaniola on his fourth and last voyage, he had again recognized the signs of an approaching storm. Columbus requested shelter in the harbor of Santo Domingo, but the new Spanish governor refused the request. That man was a friend of the former Governor Francisco de Bobadilla who was a notorious enemy of Columbus. In fact, the admiral's rival de Bobadilla himself arrogantly ignored Columbus's warning and departed for Spain with thirty ships loaded with gold and slaves.

But Columbus had been right in his storm forecast. The 1502 hurricane hit the Caribbean with incredible intensity.

Two-thirds of de Bobadilla's fleet—including the former governor himself—were lost and never seen again. All but one of the remaining treasure fleet of ships had to creep back to Santo Domingo in tattered shreds. In an ironic twist, the single ship of de Bobadilla's treasure fleet to make the Atlantic crossing back to Spain was the *Aguja*, which carried the gold that Columbus himself was owed. That coincidence spurred accusations that Columbus had magically summoned the storm out of pure spite against the former governor.

Columbus himself described the hurricane's passage: 'The storm was terrible and on that night [my other two] ships were parted from me. Each of them was certain the others were lost' (Alexander 1902, 68–69). The new capital of the Hispaniola Island, Santo Domingo, was razed to the ground. The hurricane completely destroyed forty-five wooden huts and even several stone buildings. The Spaniards, who had never seen a storm of that kind before, were perplexed by its ferocity—not least being the colony's new governor, who at great cost had to reconstruct the entire city from the ruins.

These devastating storms have long inflicted pain and destruction onto the people and places in their paths.

Hurricanes and typhoons are regional members of a general class of storms called tropical cyclones. A tropical cyclone is a circular rotating storm created by the flow of heat and moisture from the ocean into the atmosphere. Tropical cyclone size varies from as small as one hundred km (sixty miles) in diameter (Cyclone Tracy in 1974) to as massive as 1100 km (675 miles) in diameter (Typhoon Tip in 1979). Such storms are often characterized by a clear, nearly cloud-free, spot in their center, known as an *eye*. A tropical cyclone's eye is surrounded by a donut-shaped circular ring of immense thunderstorms, the *eyewall*. It is within a tropical cyclone's eyewall that the strongest winds occur.

Surprisingly, to those who have never experienced such a storm, the greatest damage caused by a typhoon or a hurricane typically *isn't* due to the winds. Instead, the most catastrophic destruction is the result of *storm surge*. A storm surge is the inundation—the widespread flooding—caused by the storm's powerful whipping winds raising water levels and, to lesser degree, the intense low pressure of a tropical storm. As I'll discuss in chapter 13, a single cyclone's storm surge can kill as many as 300,000—yes, almost a third of a million people.

Given those tragic numbers, it is not surprising that records of tropical cyclones from around the world have long been collected, archived and analyzed. In particular, a good friend of mine, Dr. Chris Landsea of the US National Hurricane Center, has made the historical documentation of tropical cyclones both his profession and his avocation. He started out as a top student of the legendary hurricane forecaster, Dr. Bill Gray at Colorado State University. After achieving his doctorate in 1994, this NOAA award-winning scientist has made many landmark discoveries about hurricanes. He is currently the chief of the Tropical Analysis and Forecast Branch at the National Hurricane Center.

Chris tells an interesting story about one of his first up-close-and-personal encounters with a monstrous hurricane.

*Figure 10.1* Two legendary tropical cyclone experts. Left, hurricane hunter Jack Parrish holding a droposonde, one of the devices dropped into hurricanes. Right, hurricane researcher and historian Dr. Chris Landsea of the US National Hurricane Center.

Source: Left photograph courtesy of NOAA Aircraft Operations Center; right photograph courtesy of Landsea.

> . . . a couple of weeks after graduate school started, there was this hurricane called Gilbert that was moving across the Caribbean, and it got to be very strong. It went right over Jamaica, east to west, caused a lot of devastation. At that point my advisor, Dr. Gray, he approached me and a couple of other grad students. He said, 'Chris, there's a hurricane out in the Caribbean, would you like to go fly?' And I was like, 'Well, sure, I'd love to.' . . . [So a couple of other grad students and I] showed up at Miami where the Aircraft Operations Center was located and basically unannounced said, 'We'd like to see if we can get aboard a flight.' Fortunately we knew a couple of the folks there and they bumped a news crew to allow myself and these two other grad students to get aboard.
>
> Turned out that Gilbert continued to intensify, and by the time we were there, it not only was a Category 5 hurricane, but it was the strongest hurricane ever observed [up to that time], 888 millibars. There was a NOVA crew onboard doing a documentary at the time. If you listen to it and see it, you'll hear in the background some whoops and yells from myself and my two co-students in the back of the plane, just can't believe the good fortune we have in getting a chance to experience this. [Unfortunately] our good fortune turned out to be horrible for Mexico because Cozumel and Cancun really got whacked pretty hard when the hurricane made landfall the next day. (Chris Landsea 2018)

I am forced to add to Landsea's story the addendum 'up to that time,' because it was a 2015 WMO extremes evaluation that dethroned Hurricane Gilbert as the strongest hurricane ever observed. But before I get into that evaluation, let's address that elephant in the room—why do we name hurricanes anyway? They're not living creatures after all!

Until the early 1950s, tropical cyclones across the world were generally labeled (if at all) by the year and the order in which they occurred during that year. For example, Hurricane '1948a' or '1948-alpha' might have been two identifiers that researchers used to designate the first tropical cyclone of 1948 in the Atlantic. That type of labeling was good for record-keeping, but it didn't provoke enough attention when used as a warning label for the general public.

One reason for using short, memorable names for these storms is that such names are quicker to communicate. Names serve to reduce possible confusion, particularly when two or more tropical storms are occurring at the same time. In the past, confusion and false rumors resulted when storm advisories broadcast from radio stations were mistaken for warnings concerning a different storm located hundreds of miles away. Information clarity is especially important in exchanging detailed storm information between hundreds of widely scattered stations, coastal bases, and ships at sea.

A final reason for naming storms is that people tend to pay more attention to a name than a number. Be honest, would you take as much notice if I said that Hurricane 289-b was approaching you than if I said that Hurricane Randy was bearing down on your house?

In the early 1950s, storms were first systematically identified using the Greek alphabet—Hurricane Alpha, Beta and so on. Then weather forecasters—they're only human, after all—began personalizing the storms by giving them the names of their girlfriends. In 1953, the United States officially began employing female names for tropical storms with the first storm of the year being given a name starting with the letter *a*, the second beginning with the letter *b* and so on. By 1978, Northern Pacific storms were identified by alternating male and female names. This procedure was then adopted in 1979 for storms in the Atlantic basin.

The World Meteorological Organization now controls the naming of tropical storms across the entire globe. For Atlantic hurricanes, there is a definitive list of alternating male and female names, which are used on a six-year rotation. The only time that there is a change to those lists is if a storm is so deadly or costly that the future use of its name for a different storm would be inappropriate. So that means, for example, there never will be another tropical cyclone named Hurricane Andrew, Hurricane Katrina or Hurricane Sandy.

Additionally, in the event that more than twenty-one named tropical cyclones (going through the entire available alphabet, some letters aren't used) occur in a season, the WMO issues a supplemental list of names.

Lists of names now exist for every ocean basin in which tropical cyclones occur—Atlantic, Pacific, Indian Ocean and around Australia. Most areas—like the Atlantic—use male and female names. However, the WMO now labels tropical cyclones in the Northern Indian Ocean using gender-neutral names, in preference to the wishes of countries in that region.

And, of course, we keep records of these storms—the biggest, the smallest, the longest-lasting and so on.

Since the beginning of the WMO Archive of Weather and Climate Extremes in 2007, we had listed—as Dr. Landsea mentioned above—that the strongest hurricane ever recorded for the Atlantic/Caribbean was Hurricane Gilbert which primarily struck Mexico in 1988.

Strongest? We measure a hurricane's intensity by how low the central pressure is—the pressure recorded in the center of the cyclone. Scientists measured Gilbert's lowest recorded pressure at an astonishing 888 millibars (26.22 inches of mercury). For the next twenty-seven years, Gilbert held the record for lowest pressure recorded in a Western Hemispheric hurricane.

Then, in 2015, a tropical cyclone developed off the west coast of Mexico that purportedly shattered that record. Enter Hurricane Patricia.

Patricia was a late-season (20–24 October 2015) major hurricane that intensified at a rate that only rarely been observed in a tropical cyclone. Within a few days, it grew to a Category 5 hurricane (on the Saffir-Simpson Hurricane Wind Scale, the ordering system that we use to rank hurricanes) over very warm waters to the south of Mexico on the *west* side of its coast.

We tend to think of hurricanes only hitting the east coast of Mexico in the Caribbean and Gulf of Mexico, but a sizable number of tropical storms also occur along the west coast of Mexico. Many of them form and then simply drift out unnoticed into the relatively uninhabited eastern Pacific Ocean. But

*Figure 10.2* Category 5 Hurricane Patricia near the Mexican coast at 2035Z, on 23 October 2015, as seen by the Suomi NPP satellite VIIRS instrument.
Source: Image courtesy of NOAA

nonetheless a few destructive West-Mexican coastal hurricanes have hit such picturesque cities as Mazatlán and Acapulco.

In this case, to the residents' good fortune, Hurricane Patricia turned north-northeastward and weakened dramatically before making landfall along a sparsely populated part of the coast of southwestern Mexico as a Category 4 hurricane. Nevertheless, during its short and intense lifetime, Patricia produced a narrow swath of severe damage and caused two confirmed deaths.

At its height, researchers measured Patricia's central pressure at 883 mb at 1733 UTC (10:30 AM Local) on 23 October. That intense low pressure, if verified by the WMO, would be the lowest sea-level pressure ever recorded for the Western Hemisphere, only slightly higher than the lowest global pressure ever measured. That record was (and remains) an astounding pressure of only 870 mb recorded in 1979's Typhoon Tip.

A critical question addressed by the WMO committee: how exactly was that pressure measured? One would think that it couldn't have been measured by a living person—it was far too dangerous to be in that monster hurricane, right?

Believe or not, the measurement *was* recorded by an actual person.

Some of the most heroic weather people on the entire planet face death daily every hurricane season by flying *into* those horrific storms. The valiant pilots and scientists of the 'hurricane hunter' planes—the aircraft that Dr. Landsea mentioned above—are the world's first-line warriors in humanity's ongoing war with tropical cyclones. Those dedicated people flew hundreds of times each year *through* tropical storms and hurricanes.

As the epitome of such death-defying meteorologists, I must mention the legendary scientist Jack Parrish who has been a hurricane hunter meteorologist since 1980 and only recently retired in 2021. I had the distinct honor of meeting—and flying with—Parrish in a project when I had the heart-pounding opportunity to fly on one of the hurricane hunter planes in the 1990s.

During one summer before the height of hurricane season, we undertook a major scientific study of a much smaller, but somewhat similar, type of storm to a tropical cyclone. That type of storm is a convective (hot air) thunderstorm and is the type associated with huge dust storms of the North American Southwest Monsoon. At that time, our scientific project was under the direction of the king of weather forecasting, Dr. Bob Maddox of the National Severe Storms Laboratory, and as part of it, we flew a hurricane hunter into the turbulent air of southwestern thunderstorms.

That flight was one of the most intense airplane journeys that I have ever had. Imagine the most turbulent roller coaster ride that you have ever been on—but lasting perhaps four or more hours! Nevertheless, I will also attest that I have never felt safer on a plane! The pilots of the hurricane hunter aircraft are some of the most skilled aviators in the world.

I learned first hand that, with every passage through a tropical cyclone, those dedicated meteorologists put their lives on the line. They collect life-saving information on a storm's wind speeds, movement and intensity.

The hurricane hunter fleet consists of scientifically equipped aircraft that play an integral role in hurricane forecasting. The US has two specialized units—one military and one civilian—both are referred to as 'hurricane hunters.' The military wing, the Air Force Reserve 53rd Weather Reconnaissance Squadron, is the world's only operational military weather reconnaissance unit, and is based at Keesler Air Force Base in Biloxi, Mississippi. They operate ten Super Hercules WC-130J aircraft.

The civilian unit, the NOAA hurricane hunters run by NOAA's Office of Marine and Aviation Operations, is based out of Lakeland, Florida. It has two Lockheed WP-3D Orion aircraft (nicknamed 'Kermit' and 'Miss Piggy') and a G-IV Gulfstream high-altitude jet capable of flying *over* hurricanes.

These hurricane hunter planes are highly instrumented flying laboratories that can take a multitude of atmospheric and radar measurements within tropical cyclones. When I flew with them, I was amazed at the variety of science devices used onboard these flying research labs to measure storms.

Regarding our WMO investigation of Hurricane Patricia, one of those instruments that those hurricane hunters use to collect the life-saving weather data is called a dropsonde. It was exactly that kind of instrument that measured Patricia's central pressure at that jaw-dropping value of 883 mb on 23 October 2015.

A dropsonde is like a weather balloon sensor, which is technically called a 'radiosonde.' A radiosonde is the type of instrument used by weather offices around the world twice a day to measure atmospheric conditions such as temperature, humidity, wind speed and pressure. *Sonde* is both the French and the German word for *probe*. Just in the United States alone, the National Weather Service launches more than two hundred radiosondes every day from weather stations around the country. The difference is that a normal radiosonde is *lifted* aloft by a helium balloon, while a dropsonde, as its name implies, is literally *dropped* from an airplane—in this case, from one of the hurricane hunter planes.

Dropsondes were first developed in the 1960s for the US Navy and Air Force specifically for hurricane reconnaissance. They are released at specified altitudes and—unpowered for flight—they fall through the hurricane down to the sea, collecting valuable, important information. During the dropsonde's descent, a GPS unit identifies its precise location as it collects data of the surrounding atmosphere and sends those data back to the aircraft by radio transmission. The descent of the dropsonde is slowed by a small parachute, which also ensures that the device stays vertically oriented as it falls to the ocean. Under most circumstances, dropsondes are released about forty-two thousand feet (12.8 km) above the ocean surface.

A dropsonde is a tube that varies in length from roughly eleven to sixteen inches (275–400 mm) long, with a diameter of about three inches (76 mm). So, it is a bit larger than the size of the cardboard paper tube for a paper towel roll. It has a weight just under a pound. As it falls, it sends temperature, humidity and pressure back to the hurricane hunter plane every 0.5 seconds, and wind speed and direction measurements back every 0.25 seconds.

It falls fast.

Dropped from forty thousand feet (12,200 meters), a dropsonde will be traveling at twenty-eight m/s (sixty-two mph) through the hurricane, such that a normal descent lasts about seven to ten minutes (based in part on how much turbulence it encounters). Roughly thirty dropsondes are released during a typical research flight, with each flight lasting about 8.5 hours—that's about one dropsonde every fifteen minutes. From 1996 to 2012, the NOAA hurricane hunters dropped 13,681 dropsondes inside hurricane eye walls and in the surrounding environment for 120 tropical cyclones.

Given the superlative quality of work by the hurricane hunter scientists and the caliber of the instrumentation, our evaluation was an easy one.

The WMO evaluation committee, having reviewed a detailed report on Patricia created by the US National Hurricane Center, verified that the dropsonde measurements had been made in a valid and approved manner. Although the hurricane had weakened by the time it made landfall, observations of wind and pressure made by land-based observers corresponded with the airborne measurements. Therefore, the committee recommended to me that we accept the measurement of Hurricane Patricia's central pressure of 883 mb at 17:33 UTC (10:30 AM Local) on 23 October 2015 as the lowest recorded pressure for the Western Hemisphere. I accepted that recommendation with pleasure.

The inclusion of Hurricane Patricia's amazing central pressure into the WMO World Archive of Climate and Weather Extremes, along with other parameters for tropical cyclone strength, intensity and distance traveled serves a continual testament to the courage, skill and daring of the many heroic meteorologists—in the air and on the ground—involved in their collection.

# Interlude: Freaks of Hurricanes

In this chapter, I mentioned my friend Chris Landsea of the US National Hurricane Center. For many years in Florida, Chris has annually hosted an interesting weather-related party in July for his friends and colleagues, patterned from a quasiofficial holiday in the US Virgin Islands called a 'Hurricane Supplication Day.' According to many Christians in the Caribbean, that day marks the beginning of the hurricane season for the region. Residents commemorate the day with special church services to pray for safety from the storms that ravage the Caribbean.

The custom dates to the rogation ceremonies which began in fifth-century England. *Rogation* derives from the word *rogare*, meaning 'to beg or supplicate.' Rogations preceded Ascension Day. By chanting the litany of the saints before Ascension Day, Christians are beseeching God for protection from calamities. On the US Virgin Islands, the fourth Monday of July is particularly designated as Hurricane Supplication Day.

The Roman Catholic Church also has long had another means of protecting the Caribbean for the ravages of hurricane. It once decreed that *Ad repellendam tempestatis* prayer should be intoned in Catholic Mass during the hurricane season.

The great nineteenth-century Cuban hurricane expert, Father Viñes (sometimes called the Hurricane Priest for his insightful pioneering research on tropical cyclones) noted:

> So ancient is the belief in this rule [that hurricane tracks are found farther to the west as the hurricane season progresses] that the ecclesiastic authority from time immemorial, wisely ordained that priests in Porto Rico should recite in the mass the prayer, 'Ad repellendat tempestrates,' during the months of August and September, but not in October, and in Cuba it should be recited in September and October, but not in August. All of which proves that the ecclesiastical authority knew by experience that the cyclones of October were much to be feared in Cuba but not those of August, and that in Porto Rico, on the contrary, the hurricanes of August are disastrous, while those of October are rare. (Garriott 1909, 929)

While we tend to pay attention—rightfully so—to human causalities in hurricanes, there are many other victims from the passage of these monstrous storms. In particular, birds often suffer at the savage winds of tropical cyclones.

In 1881, the Army's chief signal officer for Carteret County wrote in his annual report:

> Morehead City, N.C., 24th, over thirty hours in advance of the [hurricane], the skies became blackened with seabirds of every kind, size, color and description, moving rapidly towards the west, as if fleeing from the violence of the coming storm. 27th, birds slowly returning, at Newport, where the stream is very narrow, the fishes and porpoise were so wedged in that they could not move either up or down. The above incident would appear to give evidence of the possession of a wonderful instinct by birds and fishes. (Spignesi 1994, 293)

The effects of hurricanes on birds can be devastating. An authority for the American Bird Conservancy (ABC), one of the nation's leading bird conservation groups, said:

> While human safety is always the primary concern, with hurricanes, the impact such storms bring can be deadly for birds as well. Depending on circumstances, whole colonies of young birds may be wiped out — a whole breeding season gone. Parents tending young may stay with a nest and perish rather than abandon it for safer grounds. Nature can be brutal. (American Bird Conservancy 2014)

That brutality was particularly evident in a typhoon that passed by Formosa (Taiwan) in 1843. Noted hurricane historian, Dr. Cary Mock of the University of South Carolina, sent me a clipping from the *Canton Press* of a massive number of birds witnessed within a tropical cyclone's eye. The paper recorded that the *Cacique,* sailing from Chusan (Zhoushan, China):

> . . . encountered a heavy Typhoon commencing with a N.E. gale which continued with a heavy sea from the Eastward until 1 P.M., when as at Chusan, it fell suddenly calm, during which thousands of birds threw themselves on the deck. In a short time the wind rose again from the S.W. and soon increased to a terrific hurricane. (Anonymous 1843, 5)

A similar occurrence happened during Category 4 Hurricane Carla's passage over Texas in September 1961. One of those heroic hurricane hunter pilots reported that the hurricane's calm central eye was so filled with terrified birds that they dared not fly through it.

During the Cruz del Sur Hurricane of 1932, the SS *Phemius* was reportedly battered by the storm with winds over two hundred mph. The crew recalled that hundreds of birds, trying to escape from the storm, settled on the ship.

After a few hours, the birds were all dead. The captain noted that most of the birds were at death's door from exhaustion, while several had their wings broken by the hurricane winds. The *Phemius* herself suffered great damage: her funnel was carried away. She had become unmanageable, wallowing through the waves along like driftwood. Hatches were blown overboard, derricks and boats wrecked, upper and lower bridges burst.

An account by French explorer, clergyman and botanist, Jean-Baptiste Labat, from the late 1600s during his time on Martinique, is particularly evocative:

> ...we saw clouds of ramiers [pigeons], parrots, and grives [thrushes and thrashers] coming from Dominica, so beaten by hunger and fatigue that some fell in the sea, others on the sand, others in our fields, and others did not have the strength to stay on the branches where they perched upon arrival. (American Bird Conservancy 2014)

Birds—and other things—being tossed vast distances by these immense storms is a matter of record. When the destructive Category 4 hurricane named Hazel hit the coast of North Carolina in 1954, it actually dropped green coconuts, pieces of bamboo, a cup with the engraving 'Made in Haiti' and even tropical shells weighing several pounds. Given that it had traveled in a meandering path, some thousand miles between Haiti and North Carolina, it's not clear how it was able to maintain such heavy objects within it.

Deep in their airy hearts, it is likely that most hurricanes must be geographers. Many of them do like to travel.

# 11 Tornadoes, 'Nature's Ultimate Windstorms'

> We saw . . . a small whirlwind beginning in the road, and showing itself by the dust it raised and contained. . . The rest of the company stood looking after it; but my curiosity being stronger, I followed it, riding close by its side, and observed its licking up, in its progress, all the dust that was under its smaller part. . . I accompanied it about three quarters of a mile, till some limbs of dead trees, broken off by the whirl, flying about and falling near me, made me more apprehensive of danger. . . Upon my asking Colonel Tasker if such whirlwinds were common in Maryland, he answered pleasantly: 'No, not at all common, but we got this on purpose to treat Mr. Franklin.' And a very high treat it was too.
>
> Benjamin Franklin, the first US 'storm chaser' (Letter from August, 1775)

Growing up in the Great Plains, I can attest that there is something almost supernatural in watching a tornado as it spins up from underneath a supercell and starts wreaking havoc over the countryside. I have experienced that near-mystical occurrence several times in my life.

During my childhood in southeastern Nebraska, my hometown was—and still is—a hotbed of severe thunderstorm activity. My parent's hilltop home overlooked a small rural town and often, during severe thunderstorms, the sheriff of that town would drive out to our place. Together we all would scrutinize the storms for any signs of tornadoes. If we detected anything, the sheriff would radio into the police station and, within a few minutes, the town's civil defense sirens would start to bellow across the countryside. That exciting childhood memory helped to provide the incentive for me to go into meteorology as a career.

Since then, I have long had a passion for tornadoes . . . and for storm chasing.

So let's look at tornadoes and their extremes in detail.

First, this chapter's title—*Nature's ultimate windstorms*—is an homage to a fellow weather curator, one of the greatest tornado archivists of all time, Thomas P. Grazulis, who used that memorable phrase to describe tornadoes. Grazulis's mammoth thirteen-hundred-page book *Significant Tornadoes*—the 'Green Bible' as some of my students have affectionally called it, based on its distinctive emerald cover—has been a favorite research tool for my students and myself for years. In rigorous detail, that book documents more than fifty

DOI: 10.4324/9781003367956-11

thousand twisters, occurring over four hundred years within its 1,300 pages. Even more tornadoes were described in a later supplement. Part of Grazulis's success in such meticulous documentation of tornadoes across the country was his long association with the esteemed weather scientist nicknamed 'Mr. Tornado,' the incomparable Tetsuya Theodore (Ted) Fujita.

Ted Fujita was a weather researcher at the University of Chicago whose main interests were severe thunderstorms and tornadoes. Among his many accomplishments, Fujita is credited as the first person to identify a dangerous weather event called a microburst. A microburst is an intense pocket of rapidly sinking air that, upon impact with the ground, creates bomb-blast-like damage that, prior to Fujita's work, had often been mistaken for tornado damage. A few years before his death in 1998, I had the distinct honor of introducing the diminutive—and exuberant—Dr. Fujita at a weather meeting held in Phoenix. It gave me the chance to ask him about that incredible discovery of microbursts.

In his wonderful, animated style, Dr. Fujita told me that his study of microbursts had begun from his work in World War II. Back in Japan, he had studied the destruction patterns left after the atomic blasts of Hiroshima. Many storms' straight-line damage patterns in the Great Plains show similarities to those created by WWII nuclear bombs. From detailed study of those damage patterns, this brilliant scientist had deduced that the odd straight-line Great Plains damage he had observed wasn't caused by tornadoes but instead by a new kind of weather phenomenon that he called a microburst.

Fujita defined a microburst as a localized column of sinking air (an intense downdraft) within a thunderstorm that is usually less than two and half miles (four km) in diameter. Since their discovery by Fujita, we have found that microbursts can cause extensive damage, and in some instances, can even be life-threatening. For example, a microburst brought down a Delta Airlines jet at Dallas/Ft. Worth Airport back in 1985, killing 136 people onboard—and an

*Figure 11.1* Famed tornado researcher, Dr. Tetsuya Theodore (Ted) Fujita, next to his tornado wind generator at the University of Chicago.
Source: Photo Courtesy of the Hanna Holborn Gray Special Collections Research Center, University of Chicago Library.

additional person who had been driving near the airport at the time when the plane crashed. Fujita's work led to installation of airport radar technology to detect such hazardous weather, and save countless lives.

Dr. Fujita is best known for creating his scale of tornado intensity and damage, the so-called F-scale for tornadoes. That ranking system—and its updated successor, the Enhanced Fujita (EF) scale—is based on using a tornado's physical damage to estimate the vortex's winds. For example, Fujita ranked a tornado capable of blowing down small trees and uprooting bushes as an F-0 tornado (with winds less than seventy-three mph), while he classified an F-5 tornado (with winds from 261 to 318 mph) as one that is capable of creating 'incredible damage' with cars tossed hundreds, or even thousands of yards away, and homes reduced to their foundations (National Oceanic and Atmospheric Administration Storm Prediction Center 2023).

Why not directly measure the actual winds rather than estimate them afterwards from their damage?

Unfortunately for the vast majority of tornadoes occurring across the United States and the world, the necessary technology needed to estimate a tornado's winds—something called mobile Doppler radar, which I'll discuss below—hasn't been widely available. As a result, from the 1970s to the turn of the century, scientists have used *damage assessment* as the best measure of tornadic windspeeds via Fujita's F-scale. But, the superb severe weather forecaster Roger Edwards, my go-to expert on tornadoes, has urged caution using the F-scale for tornado damage:

> Tornado wind speeds are still largely inferred, or estimated; and the wind speeds on the original F scale have never been scientifically tested and proven [his emphasis]. . . Different winds may be needed to cause the same damage depending on how well-built a structure is, wind direction, wind duration, battering by flying debris, and a bunch of other factors. Also, the process of rating the damage itself is largely a judgment call—quite inconsistent and arbitrary. Even meteorologists and engineers highly experienced in damage-survey techniques often came up with different F-scale ratings for the same damage. (Edwards 2023)

I must interrupt myself a moment to say a few words about a good friend and weather icon. When Roger Edwards—a Lead Forecaster at the US Storm Prediction Center—says something, it is wise to pay close attention. That makes sense given that Roger has been interested in—and studying—severe weather for a very long time! By age six, Roger told his family and friends that one day he would be a tornado meteorologist. That early insight (and passion) led him to what might be called the nation's 'weather university'—the University of Oklahoma—where he studied under some of the best storm scientists in the business, like meteorologist Chuck Doswell.

What most impresses me about Roger Edwards is that storm weather is both his job and his passion. Take for example, when the deadly 3 May 1999

*Figure 11.2* Left, severe storms meteorologist (forecaster, researcher and storm chaser), Roger Edwards, out on one of his (many) severe weather odysseys. Right, Dr. Melissa Wagner in the field surveying tornado damage using a drone.
Source: Left photograph courtesy of Mateusz Taszarek and Edwards, right photograph courtesy of Wagner.

tornadoes hit Oklahoma. It was one of the most memorable storm clusters ever to strike the Great Plains (I'll talk about it a bit more below). Roger first put on his forecasting hat—doing his professional duty to get word out about the pending deadly tornado situation. Then, when his shift ended, he turned to storm chasing—driving towards the violent storm action with fellow forecaster Rich Thompson, and photographing eleven different tornadoes on that *one* day. Yes, that's far better than they encountered in the blockbuster *Twister*, movie fans! Then, afterwards, Roger put on his research hat—investigating the unique weather conditions of that tornado-filled day, and later writing both lead-authored and co-authored formal papers on it.

So, when Roger says to approach the tornadic F-scale with caution, I listen.

Other people have listened as well. In February of 2007, the old F-ranking of tornado damage was replaced by the Enhanced Fujita (EF) scale.

The EF scale has proven to be a much more scientific means of assessing tornado damage than Fujita's original system. While the ranking classifies tornado damage from EF-0 to EF-5 like the original, the EF system of windspeeds have been calibrated by engineers and meteorologists across twenty-eight different types of damage indicators. These indicators involve different kinds of buildings, but also a few other objects as well as trees. The EF scale considers the typical strengths and weaknesses of different types of construction. As with the original F scale, the enhanced version rates the tornado as a whole, based on most intense damage within the entire path.

In today's world, the way we now document these critical tornado tracks—and thereby estimate their winds—is often through aerial drones. One of my talented former doctoral students, Dr. Melissa Wagner, as a scientist linked to NOAA's National Severe Storms Laboratory in Oklahoma, has become a leading expert in weather-related drone use.

Dr. Wagner is currently conducting a major research effort using drones to gather data on the aftermath of severe storms. She says that by flying drones over damage affected areas, 'we are able to efficiently obtain very detailed pictures and video that can help us better estimate storm intensity and relate associated hazards with information observed in radar.' This is particularly true in sparely populated areas. Although satellite photos can help with tornado damage assessment, satellite imagery can often be obscured by cloud cover. Additionally, in most severe weather situations, urban areas have a higher priority in post-tornado damage assessment than their rural counterparts. So quick and reliable technology, such as drone platforms, is now providing a useful means of documenting tornado strength *after* the storms occur.

It might seem like I am talking about tornadoes only in relation to the United States. That is true—and that is not the result of any national bias on my part. Indeed, most of the world's tornado records in the WMO Archive of Weather and Climate Extremes are associated with North America, and particularly the United States.

No, this isn't part of some hidden world-government agenda to promote climate change—it is the result of the fact that an overwhelming number of the world's tornadoes each year *do* occur in North America, and specifically the United States. Roughly eighty percent of all tornadoes every year around the world occur within North America.

*What?!* Eighty percent?! Exactly *why* is North America being targeted by tornadoes so much more than other parts of the world? Is this the result of some super-secret evil weather weapon directed against the United States?

No, again sorry, no such weather conspiracy theories here, please.

The reason that North America experiences the vast majority of tornadoes occurring around the world is the result of the continent's unique *geography*.

Three critical ingredients for cooking up a severe thunderstorm—and thereby creating a tornado—are found within the borders of North America and in relatively few other areas of the entire planet.

In North America, cold air sweeping down from the Arctic interacts with warm moist air pushing northward from the Gulf of Mexico—and that interaction causes the warm air to be lifted aloft. Meanwhile, midway through the atmosphere, dry air from the desert Southwest is jetting towards that aerial ballet and in some—luckily rare—situations the air begins to twist and spin. A supercell thunderstorm can be created. In even still fewer cases, the internal mechanisms of some of those supercells can spawn tornadoes.

Fortunately, the explosive convergence of those three components—hot air, cold air, and dry air—occurs in few places outside of North America. Although Europe, India, China, Australia and a few other regions of the Earth do record isolated numbers of tornadoes each year, overall, there are few tornadoes recorded outside of North America.

Okay, no secret super villain plot against the United States, but most tornadoes do occur in North America, particularly in the United States. It is also true that tornadoes over the United States often do not materialize *singularly* from intense supercells, but occur in *groups*. As I have told my students when we have been out storm chasing, 'If you see one tornado, always check around for others. Particularly always look overhead. If the conditions are right to produce one tornado, they can—and often will—produce more than one. Don't let one drop down on you unexpectedly.'

And, on rare occasion, Mother Nature goes berserk, creating vast *swarms* of tornadoes over the span of a few hours.

One such Super Outbreak of tornadoes occurred in 1974. In a brief twenty-four-hour period during 3 April to 4 April, there were 148 confirmed tornadoes touching down over thirteen states and in Canada. More than three hundred people lost their lives in that horrific tornado swarm, with another five thousand injured. Of the 148 tornadoes, seven of them were ranked as F-5 using the Fujita tornado damage index. Just two of those seven F-5 tornadoes, one that hit Xenia Ohio and another that struck Brandenburg Kentucky, killed a combined total of sixty-three people. Many of those 148 tornado paths were identified and mapped by the legendary Ted Fujita, whom I discussed above.

For nearly forty years afterward, the Super Outbreak of 1974 held the record for greatest number of tornadoes occurring in a single day.

Then in 2011, the eastern United States experienced a disaster of almost biblical proportions. From 25 April–28 April 2011, a slow-moving cyclonic storm system left catastrophic destruction in its wake. A mammoth number of tornado reports were registered in the southern, midwestern, and northeastern parts of the United States. Initial reports placed the number of tornadoes as high as three hundred or more.

Soon afterwards, I was asked: could a record number of tornadoes have occurred?

The question presented a perfect problem for a new field of study called geographic information science, or what my colleagues call GIScience. Located within my own academic discipline of geography, GIScience might be considered the grandchild of cartography, the science of map-making. GIScience is concerned with the analysis, storage, visualization, and management of large—sometimes vast—amounts of geographic data. For instance, the mapping and analysis of the increasing occurrence of wildfires and other natural disasters in the western United States has become an important subject for many GIScientists and their mapping computers. As an aside, because of their usefulness for companies and governments, GIScientists have become some of the most in-demand researchers across the entire planet.

How does GIScience enter into our Super Outbreak evaluation?

After a tornado passes, usually the next day teams of NWS forecasters will comprehensively survey the resulting damage within their forecast area, but only for their forecast area. This method can be a problem in that the particularly long-lasting EF-4 and EF-5 tornadoes—sometimes called 'long-track' tornadoes—will cross from one of NWS Forecast Office's warning area into another. First, we must collect those individual track segments recorded by each of the different forecast offices in the area of the Super Outbreak. Then, we must digitally stitch them together using GIS mapping technology to establish the exact path—and number—of tornadoes.

Shortly after the 2011 Super Outbreak, at my request, a superb meteorologist, Greg Carbin, now the Chief of the Forecast Operations Branch of the NOAA NWS Weather Prediction Center, painstakingly computed the Super Outbreak's exact tornado numbers. He determined that a staggering total of 209 separate tornadoes occurred on 27 April 2011 (from midnight on the twenty-seventh to midnight on the twenty-eighth). By EF class, the outbreak ranged from sixty-two EF-0 tornadoes, seventy-seven EF-1s, thirty-four EF-2s, twenty-one EF-3s, eleven EF-4s, to four mammoth EF-5 tornadoes. Seventy-two people were killed in just one of those four EF-5 tornadoes, a long-track tornado that traveled from Hackleburg, Alabama to Huntland, Tennessee.

Since 2011, that horrific Super Outbreak continues to hold the record for the most tornadoes occurring within a twenty-four-hour period. Given the horrific death and destruction caused by Super Outbreaks, this is an extreme weather record that I hope will *not* be surpassed for a very long time.

Another WMO-listed tornado record that I am often asked about involves the widest tornado diameter ever measured.

Widest tornado means worse, right? The wider the twister, the more damage it can cause?

Well, not necessarily. Another weather friend of mine, a statistical weather wizard named Harold Brooks from NOAA's National Severe Storms Laboratory (NSSL) in Oklahoma, applied his mathematical expertise to that question of tornado damage versus size data. The esteemed Dr. Brooks, a longtime graduate of University of Illinois, joined NSSL in 1991 as a Research (and award-winning) Meteorologist. When he computed the correlation between tornado size and damage, Dr. Brooks found a statistical relationship such that, in general, wider tornadoes *do* have higher damage ratings. But—and this is an important *but*—the size or shape of any *particular* tornado doesn't link definitively to its strength. For instance, a small rope-like tornado still can cause violent damage of EF-4 or EF-5 caliber, and some very large tornadoes over a quarter-mile wide might produce only damage that is similar to that produced by EF-0 to EF-1 winds. In simple words, Dr. Brooks and other storm researchers concluded that a tornado's visual appearance can't be used to reliably judge its intensity.

Even so, people remain interested in how big a tornado can become.

When I first created the WMO Archive of Weather and Climate Extremes, we listed one particularly bad F4 tornado that struck Hallam Nebraska—not far from my childhood hometown—on 22 May 2004 as having the record widest diameter for a tornado, a vortex diameter measuring an incredible 2.5 miles (4.0 km). Surprisingly, given its immense size and strength, only one person was killed along its fifty-two-mile (84-km) path.

That record diameter for a tornado held for the next nine years.

Then, on 31 May 2013, a deadly EF-3 tornado struck near El Reno, Oklahoma. In particular, that tornado sent shockwaves through the entire meteorology community, in that tragically it killed four veteran storm chasers. Those were the first recorded storm-chasing deaths, dating back to the times of the country's first storm chaser, the legendary Benjamin

Franklin himself (see the quote at the beginning of this chapter). Over its lifetime, the El Reno tornado was responsible for eight fatalities and 151 injuries. In one of their safety videos, the National Weather Service refers to the El Reno tornado as 'the most dangerous tornado in storm observing history' (NWS Norman Oklahoma 2023).

Observational reports indicated that the El Reno tornado was extraordinarily wide. But could we officially document that width?

I stated early in this chapter that in most past tornado research, we have measured a tornado's strength by assessing the wind damage that it leaves behind, using the Fujita tornado damage scale and its successor, the Enhanced Fujita scale. But now we can accurately estimate the winds *inside* such vortices. Through revolutionary research projects like NOAA's Verification of the Origins of Rotation in Tornadoes Experiment (or VORTEX) project, started in the mid-1990s and its successors, scientists have created the means to determine the actual wind speeds within tornadoes using a technology called Doppler radar. Let's cover a few basics.

Doppler radar is a major upgrade from old standard radar that was invented back during World War II. With standard radar, we can identify exactly *where* a storm is. However, using Doppler radar, we can not only locate a storm, but we can even *look inside* that storm.

In simplest terms, Doppler radar might be thought of as a kind of x-ray machine for winds. By measuring the very slight phase shift in the radar beam's wavelength as it encounters tornado particles moving either towards or away from the radar, the actual wind speeds of the twister can be estimated. The principle is the same as the change you hear in the pitch of a train as it comes toward you compared to its moving away. That change in pitch is a modification in the wavelength of the sound created by the movement of the train. A higher pitch as the train moves towards you, a lower pitch as it moves away. With regard to Doppler radar, using the change in radar beam wavelength allows us to see the movements of air *within* a storm.

Beginning in 1995, researchers have achieved even greater accuracy and detail in studying tornadoes using Doppler radar. They have mounted their x-ray machines for storms onto vehicles. Doppler on Wheels, an imaginative project developed by the fearless meteorologist Dr. Joshua Wurman and run by the University of Illinois Urbana-Champaign and the Center for Severe Weather Research (CSWR), has attempted to get *in-situ* measurements of wind speed and tornado size. Indeed, it was one of Wurman's radar-mounted vehicles that estimated a mind-boggling windspeed of 302 mph (486 km/h) within the vortex of the Bridge Creek–Moore Oklahoma EF-5 tornado in 1999 (part of Roger Edwards's storm that I mentioned above). That is the highest assessed—but not *directly* measured (see chapter 6)—surface windspeed ever recorded for the world.

Since 1995, other groups have also created doppler-mounted vehicles, such as the University of Oklahoma's RaXPol radar unit, which uses a radar enhancement called dual-polarization to get even more accurate measurements. When the El Reno tornado struck in 2013, the Oklahoma RaXpol and the CSWR doppler radar vehicles had both been deployed. Both of the teams' radars captured valuable data in close proximity to the large tornado. Both teams computed that the tornado achieved a maximum path width of 2.6 miles (4.18 km). Ground-based damage assessments later confirmed those radar estimates. Our WMO evaluation of these groups' results, which they published in the *Bulletin of the American Meteorological Society*, concluded they were faultless.

Based on this, we at the WMO World Archive of Weather and Climate Extremes list the El Reno tornado of 2013 as having the widest recorded diameter (2.6 miles, 4.18 km) for a tornado vortex.

Even with this assessment, and to end this chapter on a cautionary note, my most knowledgeable severe storms expert, Roger Edwards, states that:

> Measuring the width of a tornado can be as messy and unclear as defining a tornado. [The] El Reno and Hallam [tornadoes] probably were close to the maximum size for tornadoes; but it is quite possible that others this size or somewhat larger have occurred that weren't sampled by high-resolution radars or surveyed so carefully in the field. (Edwards 2023)

And that principle holds true for every WMO record as well—not just tornadic extremes. As with all WMO evaluations (e.g., temperature, pressure, wind and so on), our tornado extremes are the highest extremes that have been placed before the WMO for adjudication and which have passed WMO's standards for such data. It is possible, indeed, likely, that greater extremes can and have occurred but have gone unreported. We must keep recording weather! That means only through the continual monitoring of the world's climate and weather, can we ensure that we have the best possible data for climate change analysis.

# Interlude: Freaks of Tornadoes

Tornadoes are very freaky weather phenomena and can create bizarre damage. But sometimes the destruction that they cause can open the doors to even stranger problems.

Consider, for example, the EF-2 tornado that struck Richmond Virginia in September of 2018. When that twister passed through the storied old University of Richmond campus on its seven-and-half-mile journey, its hundred-mile-per-hour-plus winds toppled a large old tree whose interior had rotted away sometime earlier. But what the university faculty and students discovered was that tree's decomposed core held a very stinging secret.

Within the decayed heart of that tornado-fallen tree was a huge hive, containing over seventy thousand very angry bees!

The University contacted Kirstin Berben, the laboratories manager in their biology department. Berben was one of the campus beekeepers and oversaw the University's apiculture program.

> 'The tree was cracked open, and it was a catastrophic situation for the bee colony inside, which I estimate was about 70,000 bees based on the amount of honeycomb,' Berben said. 'We started picking up the comb and transferring clusters of bees into a large storage bin. There was a large cluster that seemed to include the queen, so we focused on that.' (Farber 2018)

Due to Berben's quick response to the tornado's destruction, that bee colony was saved and later incorporated with the University's other hives.

Yes, tornadoes can be strange. Throughout my career, I have always found personal accounts involving tornadoes to be some of the most fascinating stories in my entire weird weather archive.

As case-in-point, an F-2 tornado struck central Indiana back in 2006. As the twister crossed Interstate 74 in Shelby County, news reports state that the trip of a young woman named Megan Mahoney came to a sudden and unexpected stop.

The tornado picked up her car off the road!

Mahoney later told reporters that she never saw the twister that struck her car—but she did hear it.

'They always say something cheesy like it sounded like a freight train,' she said. 'But that's what it sounded like.'

The reports state that Megan said she didn't recall a lot of the encounter, but she did remember the sound of objects hitting her car. Soon after that, she blacked out.

Her next memory was of walking through a nearby grassy field and saying to no one in particular, 'Hi, I'm Megan from St. Louis.'

Rescuers found that her car either rolled, cart-wheeled or had been tossed more than fifty yards across the interstate and over an adjoining frontage road before it had come to its final rest.

Mahoney was taken to Major Hospital in Shelbyville. By miraculous luck, she only suffered a mild concussion along with cuts and bruises. Based on her injuries, doctors indicated that she had been sucked out of the car through the sunroof.

'I'm still having trouble believing it,' Mahoney told reporters (National Centers for Environmental Information 2023).

As a follow-up to that story, I must mention a strange tornado story told to me personally by the legendary country music singer, Waylon Jennings. But I must hasten to add that I have no documentation of this particular event other than the country singer's own words and he didn't give me any dates for when it happened.

Jennings told me his story one autumn evening at his elegant house in Phoenix. I was there as part of a university outreach program where I had given a short presentation about weird weather—raining fish, strange lightning and such—for Jennings, his lovely wife, Jesse Colter and their guests. Although a bit starstruck, I somehow stumbled through the presentation without making too much of a fool of myself and everybody seemed to have enjoyed the show.

Afterwards, as we were chatting together, Jennings told me that he himself had experienced a tornado event that I find eerily similar to Megan Mahoney's later experience mentioned above—but one that he had never reported to anyone. He told me that once, many years ago, he had been driving down a highway in the Great Plains on a terrifically stormy afternoon when a tornado had touched down and lifted his car—and him—completely off the ground.

According to Waylon, the tornado proceeded to crazily spin his car around several times, before finally depositing the car—*without damage!*—on the other side of the road, facing back in the direction from which he just traveled.

Jennings told me that, thoroughly shaken, he began to drive . . . slowly . . . back the way he had just come. He finished his story with a throaty chuckle, saying that at the time he felt *somebody* was telling him that the trip that he had started could be postponed for a time.

Was Waylon Jennings' story true? I don't know for sure, but I do know that, at one time in the past, making a tornado forecast was forbidden by the US government!

In the Weather Bureau regulations for the year 1905 the following statement explicitly appears: 'Forecasts of tornadoes prohibited' (Wagler 1966, 23). That ban was repeated in the revised regulations of 1915 and again in 1934.

Why on earth would the government prohibit tornado forecasts? No, not for political considerations.

The ban was issued because of fear for general panic. Remember the scare created by Orson Welles' radio presentation of 'War of the Worlds' back in the 1930s?

Forecasters were instructed that they could predict local storms, but they could not explicitly mention the word *tornado* as 'it would cause public alarm and panic' (Wagler 1966, 23). At that time (compared to today), little was known about tornadoes, by both scientists and the public at large. At that time, tornadoes were regarded as the atmospheric equivalents of invading Martians—mysterious sky monsters capable of sudden and unpredictable acts of death and devastation. The governmental prohibition was ended in 1938.

But it was only in 1948 that the science had progressed to allow effective *advance* tornado warnings.

On the ominously hot morning of 25 March 1948, Air Force Captain Robert C. Miller and Major Ernest J. Fawbush issued a legendary forecast. That morning the two veteran meteorologists predicted that Tinker Air Force Base in Oklahoma was likely to be ground-zero for the occurrence of a tornado. For the prior three days, Fawbush and Miller had poured over surface and upper-air weather charts and reviewed past tornadic outbreaks. They came to the realization that there were certain similarities in the weather patterns preceding the occurrence of tornadoes across the Great Plains.

'Using our findings and incorporating those of others . . . we listed several weather parameters considered sufficient to result in significant tornadic outbreaks when all were present in a geographical area at the same time,' wrote Miller (National Oceanic and Atmospheric Administration 2023).

On that fateful morning of 25 March, they saw that most of those tornadic precursors were present, and they issued the first weather forecast explicitly mentioning the likelihood of tornadoes.

Their forecast was word-perfect. That evening a destructive F-3 tornado struck Tinker Air Force, causing considerable damage and a few injuries. But, in part due to Miller's and Fawbush's early weather forecast, there were no fatalities. The warning had worked.

A new era in meteorology had begun.

# 12 Lightning . . . and Megalightning!

Thunder is good, thunder is impressive; but it is lightning that does the work.
Mark Twain (Samuel Clemens), 'Letter to Henry W. Ruoff, 28 August 1908'

Back in 2007, when I started the WMO World Archive of Weather and Climate Extremes, I didn't expect that one of the most popular categories would be lightning. In fact, when we created the Archive, we didn't list any lightning records.

Why?

Early governmental archivists of weather records—the sources for our original set of records—didn't recognize any extremes of lightning. For early weather historians, there weren't any good means to assess, to document or—particularly—to verify lightning extremes. We needed new precise instrumentation to measure lightning—equipment that we now have!

The problem is that, in most cases, a lightning discharge is a very short-duration transitive event. It is the result of a series of complex micro-timed (millionths of a second) processes taking place within a thunderstorm—and, oh, yes, to have lightning, you must have a thunderstorm. There is no such thing as 'clear-sky lightning' – let me explain where that misconception originates.

Lightning is the product of electric charge separation. It's a static discharge of electricity that has been pumped up on nature's version of steroids. By that, I mean that lightning is the same process as the tiny spark of static electricity you see (and feel) when you walk across a carpet and touch a door handle—but on a much grander scale.

For lightning, the process begins when raindrops fall through a cloud. As they fall, they shear off electrons from near the top of the cloud and transfer them to lower parts of the cloud. Over time, this means that the top of the cloud becomes more positive in electrical charge while the bottom of the cloud becomes more negative.

With regard to electrical charges, a basic physics principle is that 'opposites attract.' As the bottom of the cloud becomes more negatively charged, the ground underneath the cloud progressively grows to be more positively charged. On the ground, one might start to notice this flow upward as their hair begins to stand on end—*not* a good sign!

DOI: 10.4324/9781003367956-12

The physics principle suggests that those two charges want to connect—but they are separated by miles of air . . . and air is actually *not* a particularly good conductor of electricity. So there must be *tremendous* electrical potential built up for electrons to flow towards the positive-charged surface below the storm. A typical lightning discharge actually involves about 300 *million* volts and about 30 *thousand* amps. In comparison, normal US household current is 120 volts and fifteen amps (in most of the world, 220 volts and fifteen amps).

A lightning discharge originates with a staggered-step movement of elections from the cloud's bottom. Each of the jagged steps, around fifty meters (165 feet) length and about the width of a pencil, is a 'stepped leader.' The zigzag path is the result of electricity attempting to flow through the most-conductive air. To humanize the physical process, the electron flow might be imagined to be a massively-long electric eel seeking the 'easiest' path down to the surface. Sometimes, two equally conductive pathways exist. The stepped leader then forks into two separate branches—giving rise to the 'tree-like' branched pattern of lightning.

For typical cloud-to-ground strikes, when the electrons are within fifty meters (165 feet) or so of the surface, a corollary flow of positive protons begins to flow upward towards the electrons. As an arcing discharge occurs between the two flows, a continuous conductive pathway forms. The channel becomes illuminated as positive charges move upward. This connection completes the electrical circuit and the 'return stroke' is underway, in which positive charges surge up through the channel. The return-stroke current only lasts tens of microseconds—as do most of the processes in a lightning discharge. Lightning is a micro-timed event.

Following the creation of the illuminated channel, another surge of electrons pours down the pathway a few microseconds later—a dart leader. Such surges continue over the course of the next few milliseconds as the electrical potential between the cloud bottom and ground is equalized. Think of the lightning column as a tiny pencil-wide drain through which an immense flow of electrons is pouring down from the charged cloud bottom. A typical cloud-to-ground flash might have three to four dart leaders surging downward to drain the cloud bottom's electrical potential. But as many as twenty-five or more dart leaders can occur in a single flash event. This is why lightning seems to flicker. Even though our eyes can't detect the micro-seconds-long existence of the individual dart leaders, we can sometimes see their strobe-like effect during the entire lightning discharge event.

As those electrons and protons flow through the channel, they produce an incredible current—upwards of 30,000 amps. That current creates enough heat in the tiny pencil-width column of air to reach temperatures of 50,000°F (~30,000°C)—and explodes that thin column of air, creating a blasting shockwave of air that we call thunder. Beyond the intense optical pulse (the flash of light), and the acoustic shockwave (thunder), the moving electrical charge also radiates electromagnetic radio fields which are detectable. A bit later, I'll discuss how those fields are of critical importance to our investigations.

Most lightning flash events are negative in character, meaning that they consist of electrons flowing away from the cloud bottom and lowering its accumulated negative charge. Occasionally, flows can originate from the cloud's upper, positive-dominated, region, and are called 'positive flashes.' Since these positive lightning flashes originate in the upper levels of a storm, the amount of air they must travel through to reach the ground is usually much greater. Therefore, electric fields associated with ground-touching *positive* lightning strikes are much stronger than those associated with negative strikes. The flash discharge's electrical potential may be up to ten times greater than their 'milder' negative lightning strikes—as much as 300,000 amps!

But with regard to WMO world records, we are most concerned with a new type of lightning event: the *megalightning* flash. A megalightning flash is defined (in part due to our extremes research) as a horizontal lightning discharge that extends hundreds of miles in length.

## Hundreds of miles?!

Most lightning flashes tend to be short-distance, short-duration events. Most last less than a second and generally travel less than thirty miles (fifty kms) from where they originate. You can prove this to yourself. Because of the speed differences between light and sound, it is easy to determine how far away a lightning discharge is. If you count the number of seconds between the flash of lightning and the sound of thunder, and then divide by five, you'll get the distance in miles to the lightning.

For example, five seconds elapsing between seeing the flash and hearing the bang means that specific lightning discharge occurred one mile away. Fifteen seconds would be three miles away and so on. If you like metric measurements, it is a one-kilometer distance for every three seconds separation between flash and bang. Of course, please keep in mind that, while counting, you should be in a safe place. Remember, if you can hear thunder, chances are that you're within striking distance of the storm. You don't want to get struck by the next flash of lightning. As the US National Weather Service advises, 'when thunder roars, go indoors.'

I should mention that movies seldom get that flash-to-bang timing right. They often had the thunder and lightning occurring at the same time—but that would mean you are at ground zero of the lightning discharge!

That time/distance relationship implies that a lightning discharge can actually hit many miles away from its origin point—sometimes far away from the storm location itself. That rare occurrence gives rise to the misconception of 'clear-sky lightning.' A lightning discharge can sometimes strike a location so far from its origin storm that, to a person on the ground, it might seem to have materialized out of thin air far from any storm.

But how do scientists measure the length and duration of lighting? Of course, because of safety concerns, we don't have students standing out in lightning storms timing the difference between thunder and lightning with stopwatches. That means, unfortunately for movie-goers, there isn't likely to be an action-filled *Megalightning* movie as a sequel to the blockbuster *Twister*. There are no lightning-chasing vehicles braving death and destruction to track lightning flashes!

What we do have—and still use—are ground-based recording stations. Remember that I said above that, besides the visible lightning flash and acoustic thunder, the moving flow of electrons also radiates *detectable* electromagnetic fields.

Have you ever noticed the static hiss that breaks up broadcasts on an old-fashioned radio during a thunderstorm? That hiss is VHF (very high frequency) electromagnetic energy that is originating from the lightning discharge. As the negative stepped leaders move downward from the cloud bottom, those electrical flows can be detected by high-frequency radio emissions as received by ground-based detection networks called Lightning Mapping Arrays (LMAs). An LMA network of lightning monitors is able, through geographical triangulation, to determine the exact location and time of the lightning discharge based on the time it takes the VHF 'hiss' to arrive at the various antenna stations.

There are many LMAs existing around the world. For example, an LMA established in Houston Texas was established in April 2012. It is a surface-based network of twelve sensors which monitor VHF radiation source events from lightning on a locally unused TV channel (such as a 'channel 3' that operates at 60–66 MHz). Those sensitive monitors permit calculation and display of lightning information around the Houston Metropolitan area and indeed over the whole network (which includes the Gulf of Mexico).

The Houston LMA is being run by Texas A&M University professor Dr. Timothy Logan. Tim is one of the talented 'new-generation' of lightning researchers. He is a graduate of the University of North Dakota and has had a great interest in supporting K-12 STEM learning. His main research interests are focused on aerosol-cloud interactions, aerosol-lightning interactions, and biomass burning smoke aerosol transport. More about his important work monitoring lightning later on.

*Figure 12.1* A trio of the new generation of lightning experts. From left to right: Dr. Michael Peterson of Sandia National Laboratories (New Mexico), Dr. Timothy Logan of Texas A&M University and Dr. Daille Zhang of the University of Maryland.

Source: Photographs courtesy of Peterson, Logan and Zhang.

Okay, we can measure lightning using ground-based networks of lightning monitors.

Let's return to 2014 and have my go-to lightning expert Ron Holle enter into our story. If Ted Fujita (mentioned in the last chapter) was nicknamed 'Mr. Tornado,' Ron must be known as 'Mr. Lightning Safety.' His work on the demographics of lightning victims and damages (one of his many accomplishments!) has made him the acknowledged world expert on lightning injury—and you'll hear more about his life-saving work in lightning protection and mortality in the next chapter.

One day in 2014, Ron mentioned to me that, in some of his various 'all-things lightning-related' professional groups, members had been discussing various lightning extremes—massive lightning discharges in both flash length and duration. In particular he directed me to a noted lightning expert named Dr. Timothy Lang. Timothy Lang is a NASA research scientist who has long been involved in creating and managing a variety of ground-based and airborne lightning instruments. One of the projects that Dr. Lang has been very instrumental in establishing is a Lightning Mapping Array, like the one in Texas but this one was called *Relampago*, which is Spanish for lightning. The project involved creating a sensitive network of lightning monitors to record severe thunderstorms striking the Pampas region of Argentina.

In 2014, LMAs, such as the ones in Argentina and Texas, were some of the best means of assessing how big a lightning discharge could be. For measuring lightning distance, LMAs can triangulate the high radio frequency emissions from lightning within their network of sensors, and using meticulous timing, researchers can then map the precise flashes as they develop to determine their lifetimes.

My first WMO committee of lightning experts in 2014, headed by NASA's knowledgeable Dr. Lang, evaluated two particular flashes. After detailed analysis of two research LMA networks in the United States and Europe that had measured the flashes—and complex discussion of how exactly to measure a lightning flash—the committee recommended certification of one megalightning discharge that extended for an incredible distance of 321 km (199 miles) as the global lightning distance extreme. They also recommended certification of a second flash that occurred over southern France which lasted continuously for 7.74 seconds over France as the global lightning duration extreme.

That first investigation has been instrumental in supporting the 'rewrite' of the definition for a lightning discharge!

One of the preferred literature sources for meteorologists is the American Meteorological Society's *Glossary of Meteorology*. That respected source of weather knowledge defines a lightning flash as 'a transient, high-current electric discharge with path lengths measured in kilometers' while it defines a lightning discharge as 'the series of electrical processes taking place within 1 second. . .' Given our work documenting lightning extremes, we now know that lightning events can (rarely) extend for *hundreds* of kilometers and can (rarely) last *many* seconds. Based on that first published work on lightning extremes, many scientists now think that definition must be rewritten.

But those first lightning records barely had a chance for the digital ink to dry in the WMO Archive of Climate and Weather Extremes before we were again asked verify still *newer* lightning extremes.

That was because, even as I was accepting those first two megalightning flashes as record-holders, lightning monitoring technology was improving by leaps and bounds—literally into space itself. Scientists were realizing that identifying megalightning using ground-based LMAs wasn't easy because such incredible lightning events were pushing existing sensor networks to their geographic limits. We can't *fully* see such long megaflashes using LMAs because the ground-based arrays are often not large enough to record the entirety of these enormous lightning discharges. For example, land-based LMAs don't extend far into the oceans to identify many lightning flashes.

So lightning scientists turned to orbiting satellites. Sensitive space-based lightning mapping sensors offer the ability to measure flash extent and duration over the entire planet.

In 2016, the latest generation of US weather satellite (called a Geostationary Operational Environmental Satellite or GOES) was launched, carrying a precision optical instrument called a 'Geostationary Lightning Mapper.' That GLM *continuously* maps all lightning activity across most of the Western Hemisphere, specifically North and South America. New lightning imagers have been developed for current and future geosynchronous missions such as China's FY-4 Lightning Mapping Imager and EUMETSAT's Meteosat Third Generation (MTG) Lightning Imager. Together, these instruments are providing near-global coverage of total lightning.

But, in 2016, the US GOES-16 GLM was the only instrument that provided complete coverage of the Americas' severe thunderstorms. Those locations have the powerful storms in which megaflashes are most likely to occur – namely, the Great Plains in North America and the La Plata basin in South America.

But there was still a problem. How could a researcher identify a specific extreme megaflash out of the vast array of hemispheric snapshots produced every day by the orbiting satellite? Remember how expansive our planet is. Detecting a single specific lightning flash, even if it is hundreds of miles long, at a given instant across the entire globe is a herculean task—and can only be accomplished using specific computer software. Luckily, I was able identify a lightning mapping expert who was up to the challenge of creating such software.

As another one of the talented 'new generation' of lightning scientists, Dr. Michael Peterson recently of Los Alamos National Laboratory is the acknowledged authority on this incredible satellite lightning mapping technology. Even as a high school student back when—in his own words—he was 'bored in study hall,' Michael derived an elaborate assembly language code for weather prediction . . . code for which he was later selected as a finalist in a Texas Instruments software development contest.

However, even with such successes, Peterson found in college—like the unorthodox student discussed in 'freaks of pressure' interlude following chapter 8—that he (in Michael's own words) 'unwittingly tormented my professors by taking the non-obvious path from problem to solution.' This 'out-of-the-box' thinking almost derailed the young genius's career when those same professors refused to support his interest in attending grad school.

Fortunately for the entire discipline of meteorology . . . and (selfishly) my own WMO evaluations, Peterson was accepted into the graduate program at the University of Utah—and performed at a top-notch academic level. Upon his graduation there, he has worked as a post-doctorate fellow at the National Center for Atmospheric Research (NCAR) in Colorado and at the University of Maryland.

Peterson's most recent position has been with Los Alamos National Laboratory where his passion has been—as always—creating new computer software. His pioneering lightning software analyzing the satellite GLM sensor was selected as one of R&D Magazine's top 100 innovations of 2020. He has not only won an R&D100 award, but also won gold in a special recognition category of the competition.

For our work, the critical aspect of Peterson's lightning software is that, using the satellite mapping data, it can identify and then geo-position lightning discharges that are hundreds of miles long. It was this software that has allowed us to discover that our first 'longest distance' and 'longest duration' lightning discharge extremes, the ones identified by LMAs in Oklahoma and France, weren't that rare.

Our most recent certified lightning extremes show how far we have progressed in the study of megalightning.

After careful analysis and discussion in 2022, I convened a blue-ribbon team of international lightning experts. The panel included Drs. Peterson and Lang—and another up-and-coming lightning scientist named Daile Zhang. Dr. Zhang is a postdoctoral associate in the Earth System Science Interdisciplinary Center at the University of Maryland. Her research has focused on lightning physics, lightning detections, and instrumentations. Those three—together with other top experts in lightning—evaluated the data and recommended to me that two new lightning discharge extremes be accepted:

- The 'longest lightning flash' distance record is a single flash which covered a horizontal distance of 768 km (477 miles) across parts of the southern United States and the Gulf of Mexico on 29 April 2020. This is equivalent to the distance between New York City and Columbus Ohio in the United States or between London and the German city of Hamburg.

We even have partial independent confirmation of that amazing event from Tim Logan's ground-based Houston LMA that I mentioned earlier. Fortunately, according to Dr. Logan, even though several of the Houston lightning sensors in his LMA had undergone major repairs just before the megaflash was detected, he had enough working sensors to confirm the satellite imagery's massive lightning discharge.

Sadly, the terrific storm that produced that record megalightning discharge did destroy property throughout Southeast Texas—and it was also responsible for several deaths. However, none of these deaths were related to the megalightning event.

- The 'greatest duration for a single lightning flash' is a flash that lasted 17.1 seconds developing through a thunderstorm stretching over Uruguay and northern Argentina on 18 June 2020. Notice that duration is almost ten seconds longer than our first lightning duration extreme recorded back in 2012! The speed in which technology has improved is mind-boggling.

New records will occur in the near future and exceed our existing records as our technology and the actual observations of these amazing megalightning flashes continue to improve. After several lightning evaluations made over the last decade, I am well-prepared to assemble the WMO evaluation teams needed to analyze and verify such new records.

Over the last decade, I have been lucky to have worked with an array of superb lightning specialists—people like those I have discussed above as well as others such as Rachel Albrecht, Walt Lyons, Stéphane Pédeboy, Joan Montanya, Sunil Pawar, and the legendary Bill Rison among many others. With their expertise, we have started from *nothing* with regards to lightning extremes to our last lightning announcement which was heard around the entire world!

Around the world? Yes, interest in our lightning extremes has been noticeable.

Our 2020 WMO press release on the last extreme lightning records I mentioned above had phenomenal outreach, potentially reaching the entire population of the planet. Clare Nullis, one of the hard-working Press Officers for the WMO, told me that 'the story was covered in more than 3,500 online news

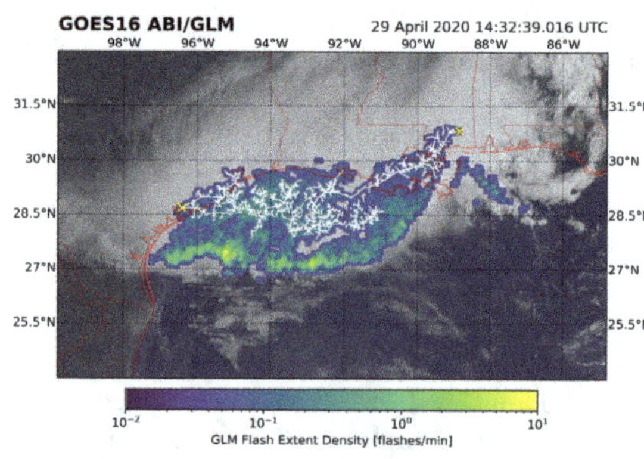


*Figure 12.2* Satellite image of record extent of lightning flash over the southern United States on 29 April 2020, which covered a horizontal distance of 768 ± 8 km (477.2 ± 5 mi). The horizontal structure (white line segments) and maximum extent (gold X symbols) of this megaflash are overlaid.

outlets and nearly 450 broadcast outlets. We had a potential reach (this tends to be an exaggeration) of 10 billion [people]—I can't ever recall such a high figure—because the story was used by [overlapping] mass circulation outlets. [Also] very widely used on social media, including tweets and retweets from main UN news account.'

Without question, I owe this amazing success of our WMO accepted lightning extremes to the lightning experts who have served on my evaluation teams. To take liberties with Mark Twain's quote from the beginning of this chapter: 'Thunder may be good and lightning may be impressive . . . but it is my talented WMO lightning experts who actually do the work.'

# Interlude: Freaks of Lightning

We at the WMO extremes project declare lightning records for only common types of lightning. But, for as much as we do know about lightning, there are many things about lightning that we have only recently discovered. There exist bizarre types of lightning that we can't yet fully explain. Of course, we don't have any official records of those 'freaky' lightning types in the WMO Archive of World Climate and Weather Extremes . . . *yet*. But maybe in the future?

One of our newest discoveries in unusual lightning has been the strange occurrence of 'dark lightning.' First discovered in 1991, dark lightning is a discharge of terrestrially-produced, very shortwave radiation called *gamma rays*—the same kind of radiation that, in the world of fiction, created the super-strong 'Incredible Hulk.' These lightning-created gamma ray flashes are some of the most powerful natural radiative emissions on the entire earth (>40 million electron volts, or MeV). Recent research has shown that dark lightning occurs—fortunately very rarely—in severe storms alongside our more normal forms of lightning.

In 2006, we gained important new information about dark lightning. On 27 October of that year, a severe thunderstorm had developed over Lake Maracaibo in northern Venezuela. Lake Maracaibo is regarded as one of the most lightning-prone regions of the entire world. Lightning can last up to nine hours during most nights of the year. One of my WMO lightning committee experts, Dr. Eldo Ávila of Argentina, noted that even as far back as 1597 the Spanish poet Lope de Vega had mentioned the famous 'Catatumbo lightning'—using the name that locals call the phenomenon. The poet has said the regional lightning had prevented the attack of English sea captain and privateer Sir Francis Drake.

One of those Catatumbo lightning events in 2006 created dark lightning . . . and that lightning was detected!

As the researchers said, 'by fortuitous coincidence,' two different lightning mapping satellites flew simultaneously over a savage Venezuelan thunderstorm on that night in 2006 (Østgaard 2013, 2423). The two independent measurements provided an unprecedented coverage of the dark lightning event.

Researchers discovered that the dark lightning's gamma rays were produced deep in the thundercloud, in the initial stages of a more normal cloud-to-cloud lightning flash, before the stepped leader had reached the cloud top and begun to extend across the cloud. Although the gamma rays of that dark lightning event were weak and short-lived (lasting seventy microseconds), a coincident radio pulse was detected at Duke University in North Carolina some 3000 km (nearly 2,000 miles) away. Researchers were able to link the North Carolina radio pulse to the Venezuelan storm—and to dark lightning.

Dark lightning's gamma radiation was detected.

Can dark lightning be dangerous? Well, sadly for the Incredible Hulk fans, gamma radiation is one of the most destructive types of radiation known to humanity and exposure to it can cause major genetic damage, even leading to death. However, dark lightning is quite rare, perhaps occurring only once for every thousand visible lightning discharges. So the moral of the story so far: please don't journey out to a severe thunderstorm in hopes of becoming a hulking green monster.

Study of dark lightning is best left to the experts.

Still another type of unusual lightning—one that has been witnessed and described for centuries—has yet to be adequately explained by science.

Ball lightning is a rare occurrence of lightning that is usually, but not always, associated with thunderstorms and severe weather. Although normally manifesting in a spherical form, it has also been observed to be ring-, rod-, or even teardrop-shaped. Sometimes ball lightning decays silently. Other times it can explode with violence. It likely contains a considerable amount of energy, since there are documented reports that it has damaged objects and even killed people. On the other hand, some people have been struck by ball lightning and survived without injury or harm.

I wish that I could tell you what ball lightning actually is.

The simple truth is that—while we have developed some interesting theories about it—we don't know.

Anecdotes about ball lightning date back to the time of the lightning pioneer Benjamin Franklin. You probably know of Franklin and his famous kite experiment proving that lightning was electricity. But you likely haven't heard of a Swedish contemporary of Franklin's, G. W. Richman.

G. W. Richman was an early experimenter working with a variation of Franklin's lightning rod that he called an 'electrical gnomon,' consisting of an ungrounded lightning rod erected upon the top of Richmann's house in St. Petersburg, Russia. In the words of meteorologist Charles Tomlinson:

> While the professor was stooping down to see the amount of electricity indicated by the electrometer, a large globe of bluish-white fire flashed from a metal conducting rod, with a report as loud as that of a pistol. The professor fell back upon a chest behind him and instantly expired. [An assistant] was stupefied and benumbed by the shock, and was struck by several fragments of red-hot wires; but the moment he recovered he ran out of the house calling for help. In the meantime, Mrs. Richmann, who heard the stroke of the thunder, hastened to the room and found her husband sitting lifeless upon the chest. The house was filled with a sulphurous vapour; a

> clock in an adjoining room was stopped; the ashes were thrown from the fireplace, and the door-posts of the house were rent asunder. Medical aid was of no avail to the professor; a red spot appeared on his forehead, and several red and blue spots were found on the body. The shoe on the left foot was burst open; and below the aperture was a blue mark on the foot from which it is probable the electricity had issued. The stocking was entire at the place where the shoe was burst, and the coat had received no damage. The back of the engraver's coat, however, was marked with several long and narrow stripes, as if red-hot wires had burned off the nap. (Tomlinson 1860, 51–52)

Was Richmann struck by ball lightning? It is impossible to know for sure but the descriptive word 'globe' in contemporary accounts of the event is given as evidence of ball lightning. Nevertheless, we can't recreate what happened to the unfortunate lightning researcher.

Theories abound for explaining ball lightning. One popular theory suggests that ball lightning is a type of plasma—a fourth state of matter comprised of free charged particles like ions or electrons. Other speculations suggest that ball lightning is the result of a failed lightning bolt or is comprised of hot ionized silica created by lightning striking sandy ground. Some, less credible, reports have suggested alien UFOs cause ball lightning. But quality scientific data on ball lightning remain scarce.

I think the respected lightning expert Martin Uman's thoughts on ball lightning may be best. He stated, 'In my opinion, there are probably multiple causes for what's described as ball lightning' (Duncombe 2021).

Hopefully, a research project conducted by a New Mexico Tech physicist and a Texas State University engineer may shed better insight into ball lightning. They are currently collecting eyewitness accounts via a website (http://kestrel.nmt.edu/~rsonnenf/BL/#REPORT) to improve the basic understanding of the phenomenon. They'll compare those anecdotal accounts with actual meteorological data from weather radar systems to determine precisely what factors could lead to ball lightning.

That means, with more data, someday we'll be able to determine the physical cause of the bizarre phenomenon known as ball lightning.

# 13 Death by Weather!

> Nature always strikes back—and it is already doing so with growing force and fury.
>
> UN Secretary-General António Guterres, 'Address to Columbia University in New York, 2 December 2020'

Death fascinates people, particularly death caused by weather. Many of us—while safely at a distance from the tragedy—are morbidly enthralled by horrific calamities that can be created by severe weather. Others of us are horrified at the tragic loss of life. But with either attitude, getting the numbers right about weather deaths is important.

Remember back in chapter 1 when I discussed that errant newscaster reporting that Hurricane Katrina was 'the worst hurricane of all time.' One of my original purposes in purposing the WMO World Archive of Weather and Climate Extremes back in 2006 was to be a source of the public—and media—to 'get the numbers right.' Over the archive's first decade of existence, that purpose had been achieved—at least with regard to physical characteristics of weather phenomena such as tornado width, lightning length and so on.

But what about weather-related mortality?

We needed an WMO evaluation and this was a particular evaluation that I decided to start on my own.

I formed an evaluation committee to establish valid weather-related mortality statistics. As a start, I tasked the committee with the evaluation of five mortality extremes:

- The highest death toll for a tropical cyclone
- The highest death toll for a tornado
- The highest death toll for a hailstorm
- The highest death toll directly caused by a single lightning flash
- The highest death toll indirectly caused by a single lightning flash

With experience gained from many past evaluations, I pulled together an incredible number of weather experts on tornadoes, tropical cyclones, hailstorms and lightning-caused mortality.

DOI: 10.4324/9781003367956-13

By its end, this investigation involved a mammoth WMO evaluation committee of twenty-one extraordinary members. Many of those are people whom I had discussed in earlier chapters, such as Dr. Pierre Bessemoulin of Météo France; Dr. Tom Peterson of the WMO; Chris Burt, our talented weather historian; Dr. Abdel Wahab of Egypt; and Dr. Blair Trewin of Australia. But I also was fortunate to have many other brilliant people serve as well.

Since we were going to examine lightning-involved mortality, I definitely needed 'Mr. Lightning Safety' Ron Holle, whom I mentioned in the last chapter involving lightning extremes. But I should note that Ron Holle's expertise goes far beyond lightning. After getting his bachelor's and master's degrees in meteorology from Florida State University, Ron worked at *four* of the top NOAA research laboratories in the United States, specifically Norman Oklahoma (severe weather and tornadoes), Boulder Colorado (hail, lightning and storms), Coral Gables Florida (hurricanes) and Silver Spring Maryland (forecasting and modeling). Most recently (until his recent retirement) Ron has been an expert scientist for the lightning detection company, Vaisala, in Tucson Arizona. He is a Renaissance meteorologist; he's done it all and over the years, I've come to value his comprehensive weather expertise.

Ron recommended that I invite *the* world authority in lightning injuries, Dr. Mary Ann Cooper, to serve on our committee—a recommendation that I could not ignore! For years, Dr. Cooper, an emergency physician by training, has been an international expert in lightning injuries and lightning injury prevention. She received her BS and MD from Michigan State University and trained in Emergency Medicine at the University of Cincinnati before going to the University of Illinois. As an early pioneer in emergency medicine, she helped to design training and accreditation standards used in emergency medicine.

Because of Dr. Cooper's landmark work treating lightning injuries, she has received numerous awards from both the medical and lightning communities, including being the first physician awarded a fellowship from the American Meteorological Society. Although she has retired from the University of Illinois,

*Figure 13.1* A trio of talented weather-mortality experts. From left to right: meteorologist Ron Holle, lightning physician Dr. Mary Ann Cooper and biometeorologist Dr. Larry Kalkstein.

Source: Photographs courtesy of Holle, Cooper and Kalkstein.

she continues to mentor young lightning scientists and doctors, research and prevent lightning injuries, and raise awareness of lightning safety.

Mary Ann is the Managing (and founding) Director of the African Centres for Lightning and Electromagnetics Network (https://ACLENet.org), a non-profit organization dedicated to reducing deaths, injuries and property damage from lightning across Africa. On a personal note: please consider visiting that website and donating to the project, if you can. You will be saving lives! As Dr. Cooper says:

> Just as we have brought down the [lightning] injury rates in the U.S., we can start decreasing the injury rates in tropical developing countries, step by step, school by school, village by village until we have an avalanche of successes.

She is a knowledgeable—and compassionate—person!

My preliminary research indicated that highest tornado mortality might have occurred in the Asian country of Bangladesh. That meant having experts on Bangladeshi tornadoes. I therefore invited Dr. Ashraf Dewan of Curtin University in Australia and I also discovered—and invited—Bangladeshi tornado expert Jonathan Finch of the National Weather Service in Dodge City, Kansas.

To address tropical cyclone mortality, I was most fortunate to accept onto the committee two respected tropical meteorologists, M. Mohapatra and D. R. Pattanaik from the India Meteorological Department.

We rounded out this very large evaluation committee with world experts on general weather mortality. These included Zhang Cunjie of the National Climate Centre in the China Meteorological Administration; Andries Kruger of the Climate Service, South African Weather Service; Tsz-cheung Lee of the Hong Kong Observatory; Rodney Martínez of Centro Internacional para la Investigación del Fenómeno del Niño in Ecuador; Andrew Tait of the National Institute of Water and Atmospheric Research Limited (NIWA) in New Zealand; Scott Sheridan of Kent State University (and who is also the editor of the *International Journal of Biometeorology*); and the esteemed Laurence (Larry) Kalkstein, Professor Emeritus in the Department of Public Health Sciences at the University of Miami.

I have to take a moment to discuss Larry Kalkstein. Dr. Kalkstein is a recognized world authority on climate-human interaction and climate-based mortality. During his long-distinguished career in biometeorology, Larry has worked in some of the most remote and bizarre locations in the world. For example, early in his career, he undertook research in the impoverished country of Burkina Faso in northcentral Africa, where he encountered the startling and repeated scene of young children leading blind men and women.

The horrific causes of that epidemic of adult blindness were tiny worms that were being spread to their human hosts by black flies which swarmed the country's few rivers. The microscopic nematodes died *in* their hosts, thereby releasing toxins that create lesions on the skin and—particularly—on the cornea of the eyes. The buildup of cornea lesions from several years of exposure to the disease caused permanent blindness. That's why, though technically known as onchocerciasis, the affliction has been more commonly called river blindness.

Dr. Kalkstein was examining the climate implications of the horrific disease. Did weather variations play a role in the infestation rates? How did drought and flood impact the occurrence of the deadly disease? His work—along with that of many others—has proven so successful that in recent years the World Health Organization has moved beyond containing this gruesome disease to a new goal of complete elimination of river blindness.

More recently, Dr. Kalkstein (although *technically* retired) has zeroed in on mitigating another deadly climate-human interaction—lethal heat waves, particularly with reference to forecasting such events in urban centers, like my own hometown of Phoenix, Arizona. His development (in cooperation with the National Weather Service) of heat wave advisories and warning has unquestionably saved lives across the United States and, indeed, across the whole world.

Twenty-one experts. I definitely had an illustrious set of investigators, as well as many thorny candidates for those experts to investigate as possible world mortality extremes.

The first question that my mortality committee addressed was how far back we should go for establishing mortality records.

Given the limited quantity of available information for early historical incidents of weather mortality, the committee recommended accepting for consideration into the WMO Archive of Weather and Climate Extremes only weather and climate extremes for the period *after* 1873. What is so special about 1873?

Remember the origins of the WMO? The year 1873 was selected as the cutoff year for accepting records because that was the year of the formation of the WMO's predecessor, the International Meteorological Organization (IMO). As I discussed in chapter 2, the International Meteorological Organization (IMO) was founded in 1873 to aid in the exchange and standardization of weather information across national borders. By limiting consideration of WMO extremes to those occurrences after the formation of the IMO, both the quantity—and the quality—of the available meteorological data around the world increased.

I accepted this resolution.

That decision allowed us to reject several pre-1873 claimants for some of the mortality extreme categories. For example, during the course of our evaluation, we had uncovered and evaluated two pre-1873 candidates for the indirect lightning death extreme. First, we identified a lightning event that occurred on 18 August 1769 at Brescia, Italy. In that event, lightning struck the eighty-foot stone tower (St. Nazaire) built into the medieval walls encircling the town of Brescia, killing thousands with most (but not all) historical sources estimating around three thousand dead.

The committee also uncovered early mention of another lightning-induced gunpowder explosion that occurred in Rhodes (in present-day Greece), 6 November [although some sources say October] 1856. In this event, lightning sparked an explosion of a gun powder depository in the Church of St. John

(Jean) on the island of Rhodes, with a purported death toll up to four thousand. My committee determined that this event had many dates cited and a wide range of casualties. For example, the number of four thousand dead was evident only in documents dated after 1905.

## A)Indirect-caused Lightning Mortality Extreme

After dismissing any pre-1873 events, our examination of high indirectly caused lightning mortality revealed several potential candidates, including the 24 December 1971 crash of LANSA Flight 508 in Peru, killing ninety-one people onboard with only one survivor. A similar event involving a lightning strike on a flight from Puerto Rico to Maryland (Pan Am Flight 214), killing all eighty-one onboard on 8 December 1963.

We determined the most appropriate candidate for this category was the tragedy that occurred near Dronka, Egypt, on 2 November 1994. During the onset of very severe thunderstorms, a flash of lightning ignited three massive oil storage tanks, each holding about five thousand tons of aircraft diesel fuel. These tanks were located on a railway line that subsequently collapsed as floodwaters built up behind it. The fuel caught fire from the lightning strike and the ensuing floodwaters swept the blazing diesel into the village, killing a very large number of people. In essence, the people of the small town were killed by an unrelenting flood of burning aviation fuel that had been ignited by lightning.

How many died? One of my WMO committee members uncovered an official document from the Egyptian Ministry of Health and Population dating to that time period that reads in part (Arabic translated to English): 'Health sector officials said that hospitals in the region had received 469 bodies from the stricken village of Dronka. Security sources said the floods caused by the storm killed [an additional] 63 people in Assiut and neighboring provinces.'

*Figure 13.2* Locations of the five weather-related highest mortality events.
Source: Cartographer: Barbara Trapido-Lurie.

Based on this information, the WMO evaluation committee recommended that I accept the 'highest mortality associated with an indirect lightning flash' as the lightning flash that killed 469 people in an oil tank fire in Dronka Egypt, on 2 November 1994.

## B)Direct-caused Lightning Mortality Extreme

The claimant for 'highest mortality associated with a *direct* strike from lightning' was the least officially documented incident of the five mortality investigations. In contrast to Dronka incident discussed above, the committee addressed 'direct strike' as referring to death resulted solely from the effects of the lightning itself.

At the beginning of the investigation, the best claimant that we identified for highest direct mortality from lightning was the lightning flash that killed twenty-one people who all had been huddled within a single hut in Manica Tribal Trust Lands in eastern Rhodesia [now Zimbabwe] on 23 December 1975. With regard to this event, the primary sources of information consisted of (a) a news release from the Reuters News Service for 24 December 1975 and (b) a listing in the *Guinness Book of World Records* for 1977. With regard to the first source, the Salt Lake City *Tribune* Newspaper for 25 December 1975 states 'Bolt of Lightning Kills 21':

> Reuters New Agency. Salisbury, Rhodesia — Lightning killed 21 people when it struck a hut in which they were seeking shelter from rain, Rhodesian police said here Wednesday. The dead included 14 children. Three people survived the incident, which occurred Tuesday in the Manica Tribal Trust Lands in eastern Rhodesia. The total number of people killed by lightning in Rhodesia since Oct. 1 is now 53 — one of the worst periods on record (12A).

After contacting people at the Thomson Reuters News Service and at Guinness Book of World Records, unfortunately little additional information was uncovered. Although evidence supporting this claimant was far from ideal, the consensus of the committee was that this record would be accepted until and if we obtain more information or uncover another event with a greater death toll.

It was the recommendation of the WMO evaluation committee that I accept the highest mortality directly associated with a single lightning flash as the lightning flash that killed twenty-one people in a hut in Manica Tribal Trust Lands in eastern Rhodesia (now Zimbabwe) on 23 December 1975.

This grim single-flash lightning extreme unfortunately underscores a horrific statistic. Nearly ninety percent of sub-Saharan buildings, especially homes, are *not* lightning safe, leaving entire families, classrooms and workers vulnerable. In particular, schools and homes of that area tend to be mud-brick with thatch or sheet metal roofs held down by rocks—and vulnerable to lightning strikes. Again, as a personal note, please consider donating to Dr. Mary Ann Cooper's nonprofit lightning-safety-education organization, African Centres for Lightning and Electromagnetics Network (https://ACLENet.org) or its counterparts in South Asia (SALNet – South Asian Lightning Network), and/or Latin America (LALENet, Latin American Lighting Education Network), to help prevent this grim lightning extreme from ever being repeated!

## C)Highest Hail-produced Mortality Extreme

The primary claimant for highest mortality associated with a hailstorm was the storm occurring near Moradabad, India, on 30 April 1888. Reports say this hail event killed as many as 246 people with hailstones as large as 'goose eggs and oranges' and cricket balls (Anonymous 1888, 420). A distant second-place event (discovered by one of my committee) in Nanking, China in 1932 reported that two hundred people were killed and thousands injured by a hailstorm that struck in Honan Province.

Early sources to the event confirm a high fatality value for the Moradabad, India, event. The London *Times* (10 May 1888) reported that:

> India has been visited by a series of phenomenal storms, partaking very much of the character of the Dacca tornado. At Moradabad 150 deaths are reported, caused chiefly by hailstones. . . In Lower Bengal, at Rayebati, 2000 huts were destroyed, while 20 persons were reported to have been killed and 200 severely injured.

The journal *Nature* reported on the same day (10 May 1888), citing the same information as given in the London *Times*.

One of earliest and most complete references to this event was given in A. W. Greely's 1888 book *American Weather*. In that book Sir John Eliot, Meteorological Reporter to the Government of India, is quoted as saying:

> A terrific storm of hail followed, breaking all the windows and glass doors. . . It was nearly dark outside, and hail-stones of an enormous size were dashed down with a force which I have never seen anything to equal. . . Two hundred and thirty deaths in all have been reported up to the present time. The total number may be safely put as under two hundred and fifty. The majority of the deaths were caused by the hail. (Greely 1888, 234)

One of our committee members has also uncovered Sir John Eliot's official daily weather observations for this time. Although Eliot's daily documents do not specifically mention a death toll, they do note 'the [weather] conditions are very abnormal today, and weather is generally disturbed over the whole of Northern and Central India.'

Most professional accounts cite or confirm the value given in the Greely book. Noted hail expert Snowden Flora in his well-recognized book *Hailstorms of the United States* stated that in the 30 April 1888 hailstorm, 230 people were killed at Moradabad and sixteen others died at Bareilly to give a combined death toll of 246 people. This figure has been also cited by other meteorologists, including C. F. Talman in his 1931 book, *The Realm of the Air*, and noted weather historian, Patrick Hughes, in the magazine *Weatherwise*.

This event falls within the time range recommended for adjuration of extremes for the WMO Archive of Weather and Climate Extremes (that is, 1873 to present). It was the unanimous recommendation of the WMO evaluation committee to accept the highest mortality associated with a hailstorm as the storm occurring near Moradabad, India on 30 April 1888, which killed 246 people with hailstones as large as 'goose eggs and oranges.'

## D)Highest Tornado Mortality Extreme

The claimant for highest mortality associated with a tornado is the tornado that destroyed the Manikganj district, Bangladesh. This tornado occurred on 26 April 1989 and left the towns of Saturia and Manikgank Sadar completely destroyed. Approximately eighty thousand people were made homeless. This violent storm injured over twelve thousand and purportedly killed a large number of people. The storm struck at around 18:30 local time (12:30 UTC). The tornado cut a long track, up to a mile wide, about fifty miles (80 km) NW and N of Dhaka. The towns of Salturia and Manikganj were leveled and about eighty thousand people were made homeless.

In an evaluation shortly after the event, researchers noted that:

> . . . the type of damage, the length and breadth of the path of its travel indicated that the intensity of the tornado was of the order of F3.5 in the Fujita scale and the corresponding wind was calculated around 338–418 km/hr.

A total area of about fifty square miles (150 sq. km.) was impacted by the tornado.

Accurate death tolls were difficult to track down but committee members (some with direct access to data for Bangladesh) located several newspapers citing growing death tolls. For example, the *New York Times* for 28 April 1989 stated (citing the Associated Press) that:

> A tornado in central Bangladesh killed 600 people, injured 12,000 and devastated more than 20 villages, the Government said today. At least 200 people were reported missing. The tornado Wednesday night blew away people, houses and animals as it whirled through the Manikganj area, 25 miles northwest of Dhaka, the capital.

Other newspapers also report on 27 and 28 April 1989 (Associated Press, *The Herald*, and a local Bangladesh newspaper) about five hundred to six hundred confirmed fatalities (or number of recovered bodies) with at least two hundred missing. Later newspaper reports on 30 April and 2 May 1989 (Bangladesh newspaper and *Chicago Tribune*) estimated an unofficial death toll reaching at least one thousand.

But, as a committee member noted, the significance of *only* a death toll of five hundred to six hundred is that the death toll associated with the infamous 1925 'Tri-State tornado' (traveling from Ellington, MO to Princeton, IN) in the United States was 695 people, 'so to accept the Bangladesh event as a world record we would need to be confident that the number of deaths was greater than that.'

Consequently, the committee scoured the available literature to ascertain the source of the commonly-quoted death toll of 1,300. The most prominent secondary professional source for the Saturia Tornado's 1300 mortality value is given by noted tornado historian Thomas P. Grazulis (remember his 'green bible' tornado compilation in chapter 11). The committee contacted Grazulis and he stated that he had obtained that dead toll value from articles in the London *Times*.

Upon exploration of that newspaper's archives, committee members located writings relating to the Saturia Tornado. A set of articles starting on 28 April 1989 and continuing to 2 May 1989 were all apparently written by the *Times* correspondent in Bangladesh, Ahmed Fazl. Of particular note, on 2 May 1989, Fazl wrote in a short *Times* article:

> Storm hits survivors. Dhaka – Thousands of survivors in the devastated town of Shaturia [Saturia] in central Bangladesh passed their fifth night in the open as a fresh storm swept away the few structures left standing after last week's tornado. Yesterday a convoy of army lorries arrived in the town with supplies to stave off starvation and epidemics. About 1,300 people are thought to have died in the tornado. Some 80,000 people have been living in the open.

As far as can be established, that value cited by Fazl is the best available mortality tally for this event.

It was the unanimous recommendation of the WMO evaluation committee that I accept the highest mortality associated with a tornado as the 26 April 1989 tornado that destroyed the Manikganj district, Bangladesh (23°50'N, 90° 5'E, elevation: nine meters [30 ft]) with an estimated death toll of 1,300 individuals.

## E)Highest Tropical Cyclone Mortality Extreme

For our mortality extremes, our last investigation was the one which prompted the creation of the WMO Archive of Weather and Climate Extremes: tropical cyclone mortality.

As I mentioned in chapter 1, our best claimant for highest mortality associated with a tropical cyclone was the Bangladesh (at time of incident, East Pakistan) Cyclone of 12–13 November 1970. This notorious tropical cyclone is sometimes referred to as the 'Great Bhola Cyclone' with an estimated three hundred thousand (low end) to five hundred thousand (high end) storm-related fatalities. Most of those deaths were the result of a large storm surge—massive coastal flooding—that overwhelmed the islands and tidal flats along the shores of the Bay of Bengal.

The origin of the Great Bhola Cyclone can be traced back to the remnant of an old tropical storm that moved westward across the Malayan Peninsula into the Bay of Bengal. That depression strengthened and likely attained storm intensity on 9 November 1970, while continuing to drift slowly northward. The highest winds estimated for the storm were one hundred knots (51.4 $ms^{-1}$ or 115 mph) in the November 1970 cyclone. That would be equivalent to a Category 2 on the Saffir-Simpson scale.

My committee raised two major concerns in the investigation of this proposed extreme. First, did the official records match the high range in death tolls (300,000-400,000) for this incident? Second, were there other tropical cyclones that could potentially have had a higher death toll?

As with any disaster of this size, exaggerated death tolls are common and official values difficult to obtain. For example, online sources and published sources often differ in precise numbers associated with specific disasters. One of the first professional articles addressing the death toll of this 1970 tropical cyclone was co-authored by Dr. Neil Frank, the legendary hurricane authority (and former Director of the US National Hurricane Center). He and a colleague stated:

> On 12 November 1970, a severe tropical cyclone of moderate strength riding the crest of high tide lashed East Pakistan with a 20-ft storm surge and killed approximately 300,000 people. The official figures show 200,000 confirmed burials and another 50,000 to 100,000 missing.

In 2012, a different branch of my own organization, the World Meteorological Organization, cited that mortality value of three hundred thousand as an official death toll. During the course of our own investigation, the members of my WMO evaluation committee have also uncovered many recent professional documents citing the 300,000-mortality value.

However, less regulated sites such as Wikipedia stated five hundred thousand people dead. Unfortunately, these sites claiming a 500,000-death-toll list weblinks as sources that are dead, inactive or, as one committee member noted, are 'part of the "grey" literature.' Given the consistency in the reviewed professional literature (particularly articles dating to times immediately after the catastrophe), the committee recommended acceptance of the estimated three hundred thousand value as best available estimate of the Great Bhola Cyclone mortality.

Before we could accept the Great Bhola Cyclone as the deadliest recorded cyclone, the question remained as to whether other cyclones might have produced equal or larger death tolls. For example, in their review of the Great

Bhola Cyclone, Frank and his colleague Husain stated that the 'Bakerganj' cyclone in 1876 killed between 100,000 and 400,000 people. But, as one committee member noted, Frank and Husain listed only the *lower* estimate in their own table of mortality by declaring a value of one hundred thousand dead for the Bakerganj cyclone. Other claimants such as the potential three hundred thousand deaths in the typhoon of 1923 for Japan have difficulties in that many of those deaths were caused by the Great Kanto (Tokyo) Earthquake, which occurred almost at the same time when the typhoon hit Japan.

In a similar fashion, online sources cite the death toll due to the Haiphong Typhoon in Vietnam in 1881 as up to three hundred thousand. We uncovered a recent study that reported this horrendous figure does not match with the population in Haiphong at the time and contains an error that mixes up the number of death and the cost of damages. Another contender that is also mentioned in online sources is the Super Typhoon Nina in China in 1975. The torrential rains of the storm caused the collapse of the Banqiao Dam, resulting in severe flooding in Henan province, China. While some online sources suggest the death toll of the event may have reached up to 230,000, more professional sources state that the flooding killed 26,000 people and another one hundred thousand died of subsequent disease and famine.

Even recent tropical cyclone mortality values are quite variable. Committee members identified mortality values for Cyclone Nargis in 2008 varying from more than 100,000 to 146,000 people with a commonly quoted range of 130,000 to 140,000. However, those death tolls of Cyclone Nargis are unlikely to affect the consensus on the Great Bhola Cyclone event with an estimated death toll of three hundred thousand people.

Finally, there were some discussions among committee members regarding the uncertainty in the primary cause and mortality figure of the Great Bengal Cyclone of 1737. While some sources list that cyclone as having a death toll comparable to the Great Bhola Cyclone of 1970, committee members with particular expertise in tropical cyclones noted that death counts from large killer cyclones in historical times, particularly before modern records, are highly uncertain. This sentiment corresponds to the recommendation that only values after 1873 be accepted into the WMO Archive of Weather and Climate Extremes.

It was the unanimous recommendation of my WMO evaluation committee to accept the highest mortality associated with a tropical cyclone as the Bangladesh (at time of incident, East Pakistan) Cyclone of 12–13 November 1970, with an official estimated death toll of three hundred thousand people.

Five different ways of extreme mass murder . . . by nature.

Mortality is a gruesome topic that many people find distasteful and upsetting. But, if we open the door to studies revealing how bad weather can be, we also will be shedding light on how such disasters can be prevented in the future.

# Interlude: Freaks of Severe Weather . . . Safety!

The preceding chapter on weather mortality reviewed some gruesome storms and horrific weather-related events that have killed hundreds of thousands of people.

Let's change pace a bit by completing our examination of weather mortality with some positive ideas. Let's review some very important weather safety ideas.

Lightning safety is critical. From 2006 through 2019, lightning in the United States killed or injured 418 people. Here are some basic safety ideas.

Our extremes research has shown that lightning can travel *very* far from where it originates. So, the following safety advice is vital: if you can hear or see lightning, you are in danger of being struck. You should seek shelter immediately. Best locations are enclosed buildings. A hard-topped metal vehicle (not convertible or tractor) with the windows closed is also safe. Sheds, picnic shelters, tents or trees do *not* protect you from lightning.

Don't use a corded phone during a lightning storm as lightning can hit telephone lines and the charge can travel through the wires. But cordless phones and cell phones are okay to use. Similarly, do not use laptops that are plugged into power during lightning storms—or even Automatic Teller Machines (ATMs) as lightning can hit the power lines leading to them. Finally, don't work with water (take a bath or wash dishes) during a lightning storm as lightning can hit the water lines leading into a house and the charge can be conducted by the water.

In basic language, 'when thunder roars, go indoors'—and then wait thirty minutes after the storm is over to head back outdoors!

For tornadoes and hail safety, the advice is the same (since both are associated with severe thunderstorms). To improve your odds against loss of life or injury, the US National Weather Service suggests following this basic rule of thumb: put as many walls as possible between you and the outside storm. In other words, the more walls that the storm must knock down to get to you, the less likely it is to actually reach you—and hurt you. That means hiding in an interior closet or bathroom (on the *lowest* floor, preferably a basement) is your best safety choice.

For tropical cyclones—like typhoons and hurricanes—the best advice by far is evacuation. Just get away from the coastal area when the storm will make landfall. Remember that most deaths from such storms are caused by the storm surge—the massive coastal flooding caused by waves whipped up by the strong winds. Don't try to be a tough man or woman and ride out the storm. It is much smarter to leave the area, as most hurricane survivors confirm in their post-storm interviews.

So even though lightning, hail, tornadoes and tropical cyclones can kill, they don't *have* to kill. That's particularly true if you take the time *before* the disaster occurs to have planned out what you need to do to survive!

Do keep in mind that even with these safety tips, the number one present-day weather-related killer is by far heat. Weather experts such as Dr. Larry Kalkstein tell us that heat is a 'silent' killer in that we don't often hear of it on the evening news (Crabtree 2012). Nevertheless, it is—and unfortunately will likely remain—the greatest of all weather elements in its deadliness. For example, extreme-heat-related trauma accounts for more than 1,300 deaths per year in the United States. A paper in the medical journal *Lancet* indicated that 356,000 deaths around the world were related to heat in 2018 alone. In that same year, there were only eighteen tornado-related deaths worldwide with ten of those in the United States and, for the Atlantic Basin, a total of 172 people died due to tropical cyclones. Yes, heat is the worst weather killer of all.

A few basic—but important—heat-related safety tips (courtesy of the Centers for Disease Control):

- Never leave children or pets alone in a closed vehicle.
- Slow down and avoid strenuous activity.
- Wear lightweight, loose-fitting, light-colored clothing.
- Drink plenty of water.

As many children (and adults) across the world have been taught, one of your best survival tips for weather-related disasters is to be prepared. Knowledge is power.

# 14 The Antarctic Region

> I am hopeful that Antarctica in its symbolic robe of white will shine forth as a continent of peace as nations working together there in the cause of science set an example of international cooperation.
>
> Admiral Richard E. Byrd, statement made during International Geophysical Year (IGY) operations in 1957, inscribed on the Byrd Memorial at McMurdo Station in Antarctica

> Great God! This is an awful place.
>
> Captain Robert Falcon Scott, diary entry for 17 January 1921

Since heroic explorers such as Amundsen, Scott and Shackleton first trod its icy expanses over a hundred years ago, we have found the Antarctic region to be a place of vast extremes. So, it should be no surprise that the Antarctic's wide-ranging array of weather has played a considerable role in our WMO World Archive of Weather and Climate Extremes. It is also unsurprising to learn that my Antarctic committees have actually rewritten the book on weather and climate extremes in that vast and unforgiving place.

But what may be surprising to discover is that the Antarctic is also a place where science intersects with politics—for the ultimate good of science and humanity! In my admittedly biased opinion, the current political jurisdiction of the Antarctic demonstrates one of the most sensible actions ever undertaken by the global political community.

By international treaty, the Antarctic region is restricted from human development. While countries may claim lands in Antarctica, those claims are ignored by the Antarctic Treaty. There is to be no exploitation of that region's resources. Instead, the bottom of the world has been reserved for the implementation of scientific study. In 1960, twelve countries ratified the Antarctica Treaty. Since that time, the treaty has been supported by many other nations. At the time of this writing in 2023, the total number of nations who have signed onto the Antarctic Treaty is fifty-four. The provisions of the treaty state:

DOI: 10.4324/9781003367956-14

- Antarctica shall be used for peaceful purposes only (Article 1),
- Freedom of scientific investigation in Antarctica and cooperation toward that end . . . shall continue (Article II), and,
- Scientific observations and results from Antarctica shall be exchanged and made freely available (Article III).

That treaty makes Antarctica unique—not only from a legal viewpoint but also a meteorological one. Administration of the southern polar area is *not* part of the United Nations' mandate. Yet records of its weather and climate are important. Can we at the World Meteorological Organization address and adjudicate them?

When I started the WMO World Archive of Weather and Climate Extremes, I had selected three broad categories of records within it: *global, hemispheric* and *continental* weather extremes. We at the WMO leave the within-country extremes to the authority of those individual nations. But, because the WMO is affiliated with the United Nations, how do we regard the Antarctic? A continent? A vaguely defined region of the world? In part, the answer depends on how one defines the *Antarctic* . . . or, more broadly, *continent*.

As a United Nations/WMO Rapporteur, I have found that the lands that I, as a trained geographer, consider geographically-proper continental borders are *not* always what the United Nations considers in its definitions of continental borders. Sometimes, quite simply, those borders do not make geographic sense.

For example, the United Nations recognizes parts of the Middle East including the state of Israel within its European administration, not, as geography might suggest, as part of the United Nations' Asian unit. In a similar fashion, Greenland—an island country that is part of the Kingdom of Denmark—is also considered by the United Nations to be a part of Europe, even though Greenland is located in the Western Hemisphere and is commonly linked by geography with North America.

In light of that type of political situation, in the WMO Extremes Archive, we must employ those sometimes-odd continental definitions, which, for the WMO, are labeled, not as continents, but as WMO Administrative Regions. This can conflict with what most people around the world consider a continent to be.

That means we in the WMO Extremes Archive have elected to give *several* climate definitions for several regions of the earth. For example, following UN protocols for WMO Region VI (Europe), we list the official extremes as a high temperature of 54°C in Tirat Tsvi (Tirat Zevi), Israel and a low temperature of -69.6°C at Klinck AWS, Greenland. But we *also* list the traditional European continental extremes, which are limited to the true geographical confines of Europe. Those extremes are a maximum temperature of 48.8°C at Siracusa Sicily, Italy and a minimum temperature of -58.1°C at Ust 'Schugor, Russia.

The situation is even more complex for the Antarctic region.

By treaty specifications, the Antarctic is 'all land and ice that occurs at or south of 60°S latitude.' Suppose a weather extreme occurs on an island at 60°S. Would that extreme be the value representing the *entire* sprawling region south of latitude sixty degrees south? How should we cite weather extremes for such a vast and diverse area?

That was the first question that one of my polar extremes evaluation committees needed to address, but I'm getting ahead of myself. First, I have to introduce some of the members who over the past years have served on several of my polar committees.

I must admit that when it comes to assembling our WMO evaluations teams, I feel a bit like Nick Fury, the fictional character from the popular Marvel superhero films. With my Antarctic evaluation committees, that feeling intensifies. Just like Fury brought together the 'Earth's Mightiest Heroes' to create the *Avengers*, I must convince many of the sharpest minds on the planet to evaluate Arctic and Antarctic extremes for the WMO. These individuals truly epitomize the phrase 'best of the best.'

Let's start with Dr. John King—my go-to expert for all things polar. As a student in theoretical physics, Dr. King first tackled some of the thorniest problems in meteorology, and his work led to a job as a research scientist at the UK Met Office. But during his tenure at the Met Office, Dr. King happened to hear that the British Antarctic Survey was looking for someone to start a program of Antarctic weather studies. And he found his passion!

The British Antarctic Survey (BAS) is a storied organization within the United Kingdom's Natural Environment Research Council. It is tasked with nothing less than developing and accomplishing world-leading interdisciplinary research for the world's polar regions. The BAS makes the UK a principal player in Antarctic affairs.

*Figure 14.1* A trio of talented polar scientists. From Left to Right: Dr. John King of the Antarctic Research Survey, Dr. Susan Solomon of the Massachusetts Institute of Technology and Dr. Matthew Lazzara of the University of Wisconsin-Madison and Madison Area Technical College.

Source: Photographs courtesy of King, Solomon and Lazzara.

In my (and many others') opinion, the BAS owes a great amount of credit for that success to Dr. John King. From his very first trip in 1984 'onto the Ice,' as we who have made that journey say, John has researched—often in-person—precisely how the Antarctic atmosphere links to the polar oceans, sea ice and ice sheets. That landmark work over several decades led him to publish, together with fellow BAS colleague John Turner, what most scientists consider *the* definitive textbook on Antarctic meteorology and climatology.

But particularly in the Antarctic, Dr. King told me, meteorology can be heart-pounding. He related to me that once in 1994, he:

> . . . had just left Halley on the BAS ship RRS Bransfield after spending a couple of weeks working at the base. Within a few hours of sailing, a fire broke out in the propulsion motor. Although the fire was quickly brought under control, damage to the motor left the ship without propulsion and at the mercy of the drifting icebergs that surrounded us. After several rather tense days and extraordinary efforts led by the ship's engineers the motor was partially repaired and we got underway again—and not a moment too soon as a sizeable iceberg was on a direct collision course with the ship.

As Dr. King told me, 'It's incidents like this that remind you of the power of the forces of nature in the polar regions and how far you are from help if things do go wrong.'

Over the years, Dr. King has served on several of my polar committees. He has served with some equally illustrious members.

My next polar committee member is extraordinary. Few living people have achieved such success in their scientific career that they have geographic land features named after them. Dr. Susan Solomon is one such person. In 1994, the US Advisory Committee on Antarctic Names christened two of Antarctica's geologic landmarks as 'Solomon Glacier' and 'Solomon Saddle,' in honor of her landmark research. That is in addition to her being awarded the National Medal of Science, the highest scientific award bestowed by the US government, in 1999. As one of the group chairs of the Intergovernmental Panel on Climate Change, Dr. Solomon was a co-winner of the 2007 Nobel Peace Prize. Finally, as a clear role model for what makes a great scientist, she was also inducted in 2009 into the US National Women's Hall of Fame.

Dr. Solomon is best known has been her pioneering theory explaining the formation—and periodic destruction—of the ozone hole over the Antarctic. That is in addition to her having obtained some of the first documented chemical measurements establishing manmade chlorofluorocarbons (CFCs) as the cause of that destruction. With an impeccable academic pedigree, culminating in a doctorate from the University of California at Berkeley, she began her postacademic work studying ozone in 1981 at the National Oceanic and Atmospheric Administration in Boulder, Colorado. That work resulted in two trips onto the Antarctic Ice. The first trip, in the brutal winter of August 1986, was to meticulously study the newly discovered ozone hole as it formed. With that information and more data from a second expedition in 1987, Dr. Solomon and her team were able to gather enough data to confirm that the breakdown of ozone was in part caused by the presence of manmade CFCs in the atmosphere.

Dr. Solomon is now the Ellen Swallow Richards Professor of Atmospheric Chemistry and Climate Science at the Massachusetts Institute of Technology. She is also the author of a favorite book of mine about the legendary Captain Robert Falcon Scott and the role that weather played in his tragic demise in the Antarctic.

Remember that each evaluation committee needs a local representative—the person who has direct access to the weather data of the location. Are there such people for the unpopulated Antarctic?

For that inaccessible region, I have been able to call—frequently—upon Dr. Matthew Lazzara, a senior scientist/meteorologist at the Antarctic Meteorological Research and Data Center in the University of Wisconsin-Madison and Madison Area Technical College. As a talented student, Matthew achieved his Master's and doctoral degrees from the UW–Madison with his dissertation focused on the strange ghostlike phenomenon of Antarctic fog, an eerie weather experience that I myself have experienced on the Ice. As a side note unknown to me until recently—*small world!*—Dr. Solomon, whom I mentioned above, was one of Matthew's professors on his doctoral committee.

In simplest terms, Matthew is one of the world's top gurus for anything regarding the Antarctic's remote weather stations. That is, in part, due to his being the Principal Investigator of the Antarctic Automatic Weather Station Program. Through that program, Dr. Lazzara maintains and archives vast amounts of Antarctic weather data, including satellite composite images and the data collected by Automatic Weather Station (AWS) network in Antarctica.

But what about the staffed stations in the Antarctic? Who do I contact for those data?

That question was put to test in 2015 when an Argentine station, located on one of the islands comprising the Antarctic peninsula, experienced an unprecedented heat wave. At the Argentine research base named Esperanza, an observation of 17.5°C (63.5°F) was recorded on 24 March 2015. Luckily, I was able to contact Ms. Maria de Los Milagros Skansi of the Departamento Climatología for Servicio Meteorológico Nacional (the Argentine national weather service). She supplied me with an impressive amount of critical information—raw data, photographs and even conversations from some of the people at Esperanza.

Was that 2015 Esperanza observation the hottest recorded temperature in the Antarctic?

My team first needed to define the 'Antarctic region.' After much discussion, that initial WMO Antarctic temperature extremes committee recommended to me that the 'Antarctic region' be defined in three different ways.

First, they recommended that we at the WMO Archive define the Antarctic following the Antarctic Treaty as 'all of the land and ice shelves south of 60°S latitude.'

Using that definition, the Esperanza temperature was *not* the hottest recorded temperature for the Antarctic. Instead, a temperature of 19.8°C at the British research station on Signy Island is recognized as the hottest temperature for the official 'Antarctic region.' Signy Island is located at 60°43′S just within the Antarctic's official limits.

But the committee and I realized that acceptance of an island observation measured so close to the Antarctic official northern limits (and so far away from the actual Antarctic continent) does not match the common geographic concept of the Antarctic as a single landmass. We felt that one important goal of that investigation should be to improve general education about the Antarctic's distinct climatic regimes. For example, most of the world's public are surprised to learn how mild some parts of the Antarctic can be.

To account for this idea, besides the official Antarctic extremes, that committee recommended the creation of two new *subregional* temperature extremes: a) the highest temperature recorded on, or immediately adjacent to, the Antarctic continent and b) the highest temperature recorded at or above 2,500 meters or 8,300 feet (the interior plateau region of the Antarctic).

That meant that the 2015 Esperanza temperature of 17.5°C (63.5°F) was, at that time, the highest temperature recorded for *the Antarctic continent.*

For our third category, very high Antarctic Plateau—and, yes, most of the Antarctic is an extremely high elevation—the committee evaluated the evidence and recommended acceptance for highest temperature recorded at or above 2,500 meters (8,200 feet) an observation made by an automated station located in the interior of the Adélie Coast. That station called D-80 AWS at an elevation of 2,500 meters recorded a warmest plateau observation of −7.0°C (19.4°F) on 28 December 1989.

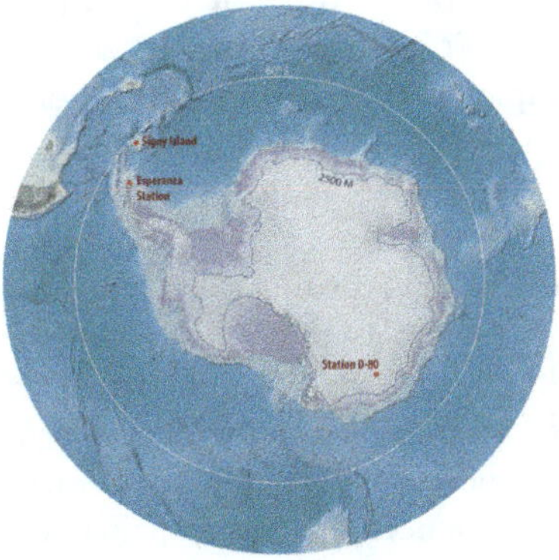


*Figure 14.2* Map displaying the three current Antarctic record high temperature locations: (a) the Antarctic region from 60°S, Signy Island; (b) the Antarctic region (continental), Esperanza Station; and (c) the Antarctic region (above 2500 meters), Automated Weather Station D-80. (Cartographer: Barbara Trapido-Lurie.)

Source: Photograph courtesy of the World Meteorological Organization.

Then, a few years later, new Antarctic records were supposedly set. But that new evaluation had a bit of excitement.

When a massive heatwave hit the peninsular region of Antarctica in February of 2020, two different locations purportedly hit record-high temperatures. The first was our 2015 old friend, the Argentine Esperanza Base mentioned above. Observers at that base recorded a new record temperature of 18.3°C value for 15:00 UTC (12 Noon Local) on 6 February 2020. Then, a few days later on 9 February 2020, an automated permafrost monitoring station on nearby Seymour Island reported an even higher observation of 20.8°C. That second station was part of a network operated by a Brazilian polar research team

My Antarctic evaluation committee thoroughly evaluated the Esperanza data, equipment and procedures. Upon completion of that process, they announced that they didn't have any troubling concerns about that observation—everything had been done by the book. They recommended acceptance of the 2020 Esperanza reading as the new highest temperature recorded for the Antarctic continent.

But would that new record last very long? Remember just a few days later, a second claimant to hottest recorded Antarctic temperature had come forth.

We had tracked down the Brazilian research team whose automated weather station had recorded—a few days after the Esperanza reading—an even higher temperature of 20.8°C. Over social media, some concerns raised about the automated weather station that had recorded the observation. To my great fortune, Professors Marcio Rocha Francelino and Carlos Schaefer of the Universidade Federal de Vicosa in Brazil stepped up to the plate in supplying us with valuable information and data, and their information allowed us to knock this evaluation out of the ballpark.

Before I continue, I must note that the Brazilian Antarctic project which recorded that 20.8°C extreme *hadn't* been designed primarily to record surface temperature. Their weather station was one of twenty-eight sites in a network of the Brazilian Antarctic Program to study permafrost. Permafrost is any ground that remains completely frozen—32°F (0°C) or colder—for at least two years straight. In these times of a warming climate, understanding precisely how the Antarctic permafrost has changed—and *is* changing—is a crucial scientific research effort.

Although the Brazilians were recording the permafrost soil temperatures and other measurements with their network of stations, they were also measuring air temperatures as well. The specific equipment that they used for such observations was appropriate: a series of temperature sensors connected to a datalogger (a device to store and transmit the environmental information), which was powered by a battery connected to solar panels. The datalogger was connected to a modem that transmitted data via satellite.

But problems arose when we obtained photographs of that one specific weather station. We saw indications that serious modifications had been made to the weather station. The air temperature sensor had been installed next to an improvised—rather than a standard—radiation shield. A radiation shield acts to keep to the station's equipment from heating up due to the direct absorption of sunlight. Think of touching a metal plate that has been left exposed in the sun—the plate will be much hotter than the surroundings due to the direct absorption of sunlight. The shield on the Antarctic weather station appeared to have been makeshift—and therefore it might not work as designed. If that were true, the temperature readings wouldn't be accurate. That is why one of the WMO guidelines is that a temperature reading must be recorded in a properly sheltered location.

I contacted Professors Marcio Rocha Francelino and Carlos Schaefer and invited them onto our evaluation committee. Both Brazilian polar scientists graciously accepted.

At this point, I must doff my hat to these incredible professors. They were both responsive and prompt in addressing this issue, going far beyond the call of duty. Without any hesitation, they acknowledged that the radiation sensor *had* been modified due to the fact that the research team didn't have the precisely specified equipment at the time of installation. A critical piece had apparently been misplaced in transit. What did that team do?

As is the case in the remote Antarctic with no handy local hardware store around the corner, the researchers improvised a workable solution to make sure that the station's datalogger, that critical device that was recording those important soil temperatures, was adequately shielded and would not overheat. Their makeshift solution worked; their key *permafrost* measurements were taken without any problems.

But did that makeshift shielding impact the *air* temperature measurements recorded next to it?

The Brazilian polar scientists offered to conduct a side-by-side set of measurements to see if the improvised shielding would cause problems. Unfortunately, like our 2010 Libya evaluation, world events once again played a role. The global Covid-19 pandemic caused the abrupt cancellation of the Brazilian research activities.

Even so, my incredible Brazilian evaluation members again came up with a brilliant workaround. They made a formal arrangement with the Brazilian Navy for military personnel to conduct a test. The Navy installed a monitoring system that had *both* conventional and improvised (for example, the kind used on the Antarctic permafrost station) sensor protection shields. The military then ran dual tests to see what difference the improvised shield caused to the temperature readings.

The results were clear and unambiguous. The improvised radiation shield produced errors of up to +5°C in daytime temperatures. The average difference between the improvised and conventional shields was +1.49°C.

From that evidence, the committee determined that the improvised radiation shield likely created a thermal bias—caused hotter temperatures—in the measurements for that one Antarctic station. They recommended to me that the 20.8°C observation on 9 February 2020 for the Brazilian Seymour Island permafrost monitoring site be rejected as a new Antarctic region extreme.

I am especially proud of our Brazilian researchers who forthrightly addressed the possibility of error in their measurements. Their work is a clear statement of the way that scientific problems *should* be addressed.

But what is the underlying cause of this flurry of recent hot temperatures in the Antarctic?

As I have mentioned several times, the world is rapidly heating up—a fact, not an opinion. That is the long-term factor in that region's warm up.

The short-term weather cause of the extremes at the Signy, Esperanza, D-80 and even the Brazilian permafrost monitoring stations was the occurrence of warm air advection—an influx of exceptionally hot air at those locations from warmer equatorward places.

Also, at the Signy and Esperanza stations, warming occurring from what are called 'foehn winds' contributed to the record high temperatures.

Foehn winds are created as air sinks down a mountainside. As that air sinks, it compresses, dries out and, most importantly, warms—sometimes in dramatic fashion. In the Appalachian Mountains of the eastern United States, for example, foehn winds have produced temperature increases of up to ten degrees Fahrenheit (12°C). In Havre, Montana, a foehn wind off the Rockies once raised the temperature from -12° to +5°C (a total of 30 degrees Fahrenheit) in just three minutes.

Foehn winds occur in many places across the world. Indeed, the name 'foehn' was first used in Roman days in the European Alps and referred to a warm west wind. In North America, a foehn wind is called a 'chinook.' As with all foehn-type winds, their extraordinary warmth and dryness result from descending air over mountains. A chinook brings relief from the cold of winter, but its most important effect is to melt snow. During a chinook event, a foot or more of snow may disappear in a few hours. This effect gives rise to a chinook's common nickname: 'snow eater.' The term *chinook* originates from a location, where such events were first recorded, a place along the lower Columbia River, the home of the Chinook people.

To give you a sense of how bizarre—and dangerous—downslope winds can be, Dr. Susan Solomon related to me a heart-pounding experience that she termed one of the most frightening in her life. It involves an incident when she was making precise spectroscopic moonlight and sunlight measurements of chemicals that drive the ozone hole. In particular, those measurements were made in part using a set of delicate mirrors on the roof of a laboratory building near the McMurdo research station.

> One memorable night I was collecting lunar data by myself, after having also taken solar observations most of the day. After several successful hours of measurements in calm conditions, the moon went down around 2 am and I wearily tucked myself into a sleeping bag on the floor of the laboratory. I figured that after a few hours badly needed rest, I would then retrieve the mirrors, as we usually did between data-taking.
>
> I was abruptly awakened by a howling katabatic [a foehn-style downslope] wind. I lay in my bag feeling the building shake. If I didn't climb the ladder up

> to the roof and get them, those mirrors would probably be destroyed and our project in shreds along with it. As soon as I got onto the roof, the wind blasted me so hard that I realized I was in danger of getting blown off. I dropped to my hands and knees, shaking with fear. I could see that the mirrors were thus far still intact, so I carefully crawled over to them. One blessing of the katabatic [wind] was that the temperatures were warm enough (around +10°F) that as I ungloved my fingers there was no brutal sting of cold to add to my struggle. I unfastened the mirrors, cradled them in their carrybag, slipped on my gloves, and slowly crawled back to the ladder. Only once did the wind blow so hard that I flattened myself altogether to avoid sliding in its powerful grip. I climbed slowly down to the snow-covered landscape and then back inside the building. I have never been so relieved to shut a door behind myself.

The study of foehn winds in the Antarctica remains a relatively new field of study. Our extremes work—linking foehn winds to extreme high temperatures—is leading more researchers to examine long-term trends in warm advection, foehn winds and extremes. Even now questions are being asked (and studied) whether foehn events are getting warmer and generating new temperature extremes.

This Antarctic investigation also led to what one of my committee members called an interesting teachable moment in media relations that I'll discuss in the next chapter.

# Interlude: Freaks of the Antarctic

Over the last few years, I have appeared as a weather and geography expert on several television shows, such as the Science Channel's *What on Earth?* and the Weather Channel's *Strangest Weather on Earth*, to discuss weird weather. These shows highlight weird weather and environments around the world in a fun and informative manner. And, without question, some of the weirdest weather and environments are found in the Antarctic.

For example, one story that I have discussed in those shows involves the McMurdo Dry Valleys, a series of strange ice-free valleys on the westside of McMurdo Sound (across from Ross Island and the US base of McMurdo Station). Flowing from one of the few glaciers there, scientists discovered a few years ago a bright crimson liquid pouring out of Taylor Glacier into Lake Bonney. Eerily, it resembled blood oozing from a wound in the ice. What on earth could such a macabre outflow be?

Scientists have discovered the cause of blood-like flow has to do with the geological trapping of water far under the ice. The water that feeds the gruesome Blood Falls was once a salty lake that is now completely cut off from the atmosphere due to the formation of thick, icy glaciers on top of the lake. Over thousands of years, the pristine liquid water under the ice has become saltier. It's now actually three times saltier than modern seawater and that's why it doesn't freeze although the water temperatures are at and even below 32°F (0°C)

The waters of that ice-trapped lake are also rich in iron. The waters exist in complete darkness and also without any air—any oxygen. So, as the iron-rich water seeps through an opening in the glacier, called a fissure, and comes into contact with the air, the iron-infused water oxidizes and rusts. The emerging liquid seeping out of the ice is stained a dark red—yes, a gruesome blood-like—color.

The thought of a desert usually conjures up images of hot, sandy plains, yet Antarctica is the largest desert in the world. Even though the vast continent is covered in ice, there is only a small amount of precipitation occurring annually—with several places recording less than 50 mm (two inches) of precipitation per year.

Even in such a dry arid environment, clouds can occur. In particular, rare clouds will sometimes form in the Antarctic very high up in the atmosphere beyond the troposphere where our weather occurs.

Such clouds are noctilucent clouds, or night shining clouds, and form as high as fifty miles (eighty kilometers) in the atmosphere. They are Earth's highest clouds and develop when small amounts of summertime water vapor rise up from the poles beyond the normal cap of the tropopause at six miles (ten kilometers). The water crystallizes around specks of meteor dust and are visible from November to February in the South Hemisphere, and May to August in the North Hemisphere. These thin clouds glow a bright electric bluish-white color and tend to appear around dusk or dawn. Unlike other clouds, they form so high in the atmosphere that they can continue to reflect sunlight after the sun dips below the horizon. The sunlight illuminates the clouds from below.

We are now studying such clouds in detail. NASA's new Aeronomy of Ice in the Mesosphere (AIM) satellite monitors noctilucent—sometimes called polar mesospheric—clouds. Researchers studying its imagery have found that morning rocket launches are partly responsible for the rare appearance of noctilucent clouds over northern parts of the United States.

'Space traffic plays an important role in the formation and variation of these clouds,' Michael Stevens of the Naval Research Laboratory said (Thomas and Hatfield 2022). He and his team compared the occurrence of noctilucent clouds to the timing of rocket launches. Their analysis revealed a strong correlation between the number of morning space launches and the frequency of some noctilucent clouds. In other words, the more morning satellite launches there were, the more noctilucent clouds were observed. Scientists are acknowledging that this linkage between our space activities and noctilucent clouds is important as scientists try to understand whether increases in weird, electric-blue clouds are connected to climate change, to human-related activities, or possibly both.

To end this interlude on a very hopeful note, one of the major Antarctic weather changes over last couple of decades involves Dr. Susan Solomon's landmark work involving the detection and explanation of the Antarctic ozone hole that I discussed. Her work started one of the greatest success stories concerning climate and meteorology for the entire planet.

Since Solomon's work, satellite measurements, that have been documenting that the Antarctic ozone hole, have shown the ozone hole has become noticeably smaller since the late 1990s and early 2000s. This reduction has been due to the implementation of the Montreal Protocol. That protocol was a treaty adopted in 1987 to ban the release of harmful ozone-depleting chemicals called chlorofluorocarbons, or CFCs. The Montreal Protocol remains the only international treaty ratified by every country on earth—all 198 United Nations member nations. Amendments have helped it evolve over time to meet new scientific, technical and economic developments and challenges.

The success of the Montreal Protocol has shown that people—when working together across the globe—can make our planet a better place now and for future generations. As the United Nation Environment Programme (UNEP) says, ' . . . the Montreal Protocol is considered to be one of the most successful environmental agreements of all time. What the parties to the Protocol have managed to accomplish since 1987 is unprecedented, and it continues to provide an inspiring example of what international cooperation at its best can achieve' (United Nations Environment Programme 2023).

# 15 Getting the Word Out

> Without publicity there can be no public support, and without public support every nation must decay.
>
> Benjamin Disraeli, 'Speech in the House of Commons (8 August 1871)'

As the preceding chapters have shown, the success of the World Meteorological Organization's Archive of World Weather and Climate Extremes is chiefly the result of the insight and dedication of the blue-ribbon evaluation teams who examined and assessed each new weather extreme. But that success could not have happened without the incredible support of a team working behind the scenes. I owe a major debt of gratitude to the people of the WMO Press Office in Geneva, Switzerland, and, in particular, one of their Press Officers, Clare Nullis.

Wait a minute. Scientists working *with* the press?

That surprised me when I first started this project.

Most scientists, I have found to my dismay, have little to no experience with the media. To the surprise of the general public, many scientists—perhaps due to their concentrated focus on specific research problems—tend to be introverts and shun publicity. That lack of familiarity with the press and communications often makes scientists appear stiff and—frankly—somewhat boring when they appear on media sites. Such an appearance can make it hard for those scientists to convey their research to the public. That opens the door for less reputable, but more charismatic, social media experts to steal the spotlight from critical research.

To counter this problem, many universities and scientific institutions such as the WMO have come to rely on a set of highly trained people who can liaison between the scientists and the media. I have been most fortunate to work with some of the best of the world's media relations people at the WMO. In particular, Clare Nullis of the WMO Press Office is a strategic communications expert with more than thirty years' experience in journalism and in the United Nations system. After working for fifteen years with the Associated Press news agency, she joined the World Health Organization before moving to the WMO. She has become one of my most valued

DOI: 10.4324/9781003367956-15

colleagues as the WMO Archive of Weather and Climate Extremes has become more and more popular around the world.

How is one of our announcements released to the world?

It involves an incredible amount of planning and hard work!

First, after a WMO evaluation is finished, we must draft an official press release. Nowadays, Clare and I work together over the course of a week or two to create a short one- or two-page statement that details—in non-technical language—the primary findings from one of my evaluation committees. That statement often includes a quote from myself or other members of the committee about the significance of our evaluation. Before publication, this announcement will go up to the office of Secretary-General of the WMO for approval. That office will sometimes add a quote from the Secretary-General himself about the relevance of the findings to climate and climate change policy.

Second, after the approval of Secretary-General, the one- to two-page announcement is meticulously translated into the various working languages and interested groups of the United Nations and WMO. These languages include English, French, Spanish, Chinese, Russian and Arabic among others. Also, we must prepare special optics (videos or visual maps or imagery). Social media (tweets and the other social posts) must be readied. Our own WMO Archive website must be prepared for the update.

Third, we coordinate any official announcement with the technical professional society, which is publishing our reviewed findings (such as the

*Figure 15.1* Ms. Clare Nullis of the World Meteorological Organization's Press Office and Dr. Melinda Shimizu, State University of New York (SUNY Cortland) geographer and lead author on a 'social dispersion' study of the WMO Extremes press release.

Source: Photographs courtesy of Nullis and Shimizu.

American Meteorological Society or the American Geophysical Union). Oftentimes, announcements must be timed so that they are released by the technical professional society at the same instant that the formal announcement is made from the World Meteorological Organization's Headquarters in Geneva, Switzerland.

Fourth, I must prepare my evaluation team of scientists across the globe for the potential media onslaught. Ofttimes, when a media outlet identifies in the press release the name of a scientist from *their* specific country, the reporters will contact that individual scientist. I want to make sure that my people are ready for that interview, particularly those who might hail from less-media-savvy areas. Often nowadays, many of my scientists belong to well-known research and governmental institutions that want to release an announcement themselves to highlight their own scientists' efforts. So we must interface with those press officials to ensure proper timing and correct release of the announcement is made.

Fifth, sometimes we need to inform the people in a place named or associated with a press release of the upcoming announcement. We employed this procedure, for example, when we released the Libya temperature announcement (discussed in chapters 4–5). By disallowing the 1922 Libyan 58°C temperature, we were elevating the 1913 Death Valley temperature to the highest recorded world temperature extreme. Therefore, before the official announcement, we interacted with the US National Park Service—the group in charge of the Death Valley National Park where that 1913 observation had been made—so that they wouldn't be caught unexpected by the announcement. And, yes, that advance notice was necessary. The Park Service told me that, after our WMO press release, they were immediately contacted by members of the media.

Sixth, I will often drop a 'heads-up' teaser to some of my weather media colleagues around the world to alert them to the forthcoming announcement. These people include such fantastic reporters as Doyle Rice at *USA Today*, Jason Samenow and the esteemed 'Capitol Weather Gang' at the Washington *Post*, and the legendary Dr. Marshall Shepherd of *Forbes*, as well as members of various television networks.

This notification serves a couple of purposes. First, from the media's perspective, it allows time for these groups to write their pieces so they release their news stories at the time of announcement. Timing is everything with the media. Second, from our perspective, it improves the visibility of the story when the WMO makes the announcement. In other words, the likelihood of a news story being picked up by other media is improved by having that story appear in other media outlets.

Then comes the formal announcement!

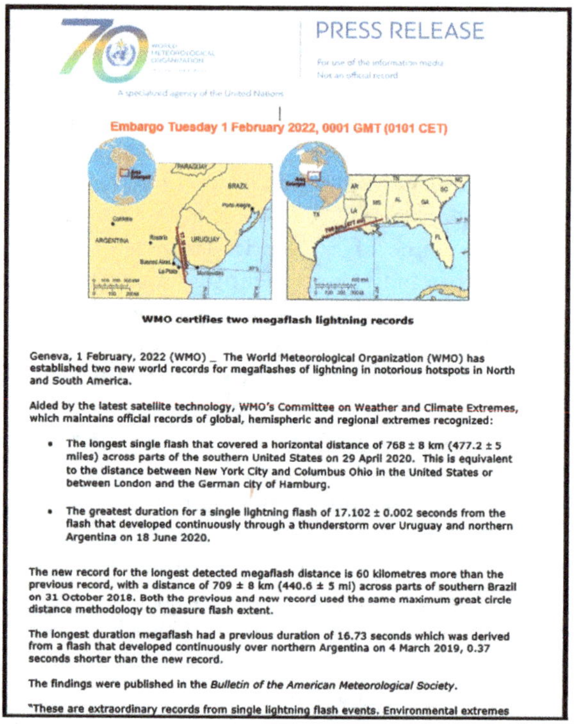

PRESS RELEASE

For use of the information media
Not an official record

A specialized agency of the United Nations

Embargo Tuesday 1 February 2022, 0001 GMT (0101 CET)

**WMO certifies two megaflash lightning records**

Geneva, 1 February, 2022 (WMO) _ The World Meteorological Organization (WMO) has established two new world records for megaflashes of lightning in notorious hotspots in North and South America.

Aided by the latest satellite technology, WMO's Committee on Weather and Climate Extremes, which maintains official records of global, hemispheric and regional extremes recognized:

- The longest single flash that covered a horizontal distance of 768 ± 8 km (477.2 ± 5 miles) across parts of the southern United States on 29 April 2020. This is equivalent to the distance between New York City and Columbus Ohio in the United States or between London and the German city of Hamburg.
- The greatest duration for a single lightning flash of 17.102 ± 0.002 seconds from the flash that developed continuously through a thunderstorm over Uruguay and northern Argentina on 18 June 2020.

The new record for the longest detected megaflash distance is 60 kilometres more than the previous record, with a distance of 709 ± 8 km (440.6 ± 5 mi) across parts of southern Brazil on 31 October 2018. Both the previous and new record used the same maximum great circle distance methodology to measure flash extent.

The longest duration megaflash had a previous duration of 16.73 seconds which was derived from a flash that developed continuously over northern Argentina on 4 March 2019, 0.37 seconds shorter than the new record.

The findings were published in the *Bulletin of the American Meteorological Society*.

"These are extraordinary records from single lightning flash events. Environmental extremes

*Figure 15.2* An example of a portion of a WMO extremes press release.

The United Nations Information Service in Geneva hosts many regular press briefings dealing with such topics as refugees, human rights, health, peace negotiations—and weather and climate. Our formal announcements are released during one of those weekly press briefings in Switzerland. As Ms. Nullis says:

> We are on a podium or desk at the front of the room and journalists in front of us. At least pre-covid. Nowadays it's hybrid and so the spokespeople still sit in the room because of UNTV [United Nations Television] filming us but most journalists follow online.

She says that many of these briefings are quite low-key and informal because it is a small group of the world's top journalists—perhaps twenty or thirty at most—attending the regular briefings. But the announcements do pack a punch! Journalists from all over the world, including the major news agencies, participate. The briefings are broadcast by UN Television to TV outlets around the world. Most of the big news outlets are there, for example Reuters, AP, AFP (French and English, with big reach in Asia and Africa), DPA (German), EFE (Spanish), Xinhua, China Central TV and the various Japanese, Brazilian and Arab-speaking outlets.

This means that announcements of new records in the Weather and Climate Extremes Archive appear in news outlets and TV channels on every single continent and in multiple languages. This coverage reached a crescendo early in 2022 when we announced two new lightning extremes (see chapter 12 for details on that actual investigation).

After the WMO had made the formal announcement, Clare discussed the astonishing media outreach of that lightning announcement:

> The coverage . . . received phenomenal reach. The story was covered in more than 3,500 online news outlets and nearly 450 broadcast outlets (including interviews with BBC, CBS, Fox). We had a potential reach (this tends to be an exaggeration because of overlapping coverage) of 10 billion [people]—I can't ever recall such a high figure—because the story was used by mass circulation outlets. Very widely used on social media, including tweets and retweets from main UN news account.
>
> [There was] huge visibility in US and Spanish-language media. Also in most of the major TV and news outlets in Europe and Japan. What was particularly striking was the unusually high uptake in developing countries. I've never seen one of our stories get so much coverage in India - including in major Hindi news outlets. BBC Kiswahili (reach of 63 million) covered it, as did other foreign-language BBC and CNN outlets. Many thanks to translators for all their help in translating the press release. . . And very special thanks to . . . the Weather and Climate Extremes Archive for the amazing work on what has become one of WMO's flagship products. We hope it helped sensitize public opinion to this major hazard.

As Clare noted, such publicity from one of our evaluations serves to inform and emphasize the importance of weather to people across the globe. But I hope that they also help other scientists to see how to communicate their own research findings to the world.

Putting that hope into action, a few years ago Melinda Shimizu, then a doctoral student of ASU's Vice-Provost and Dean Dr. Elizabeth Wentz, spearheaded an interesting geographic dispersion analysis of one of our early WMO announcements. She (with the help of her graduate supervisory committee) undertook a detailed media examination of the WMO's evaluation of the 1922 Libyan 58°C temperature (see chapters 4 and 5). I should note that Dr. Shimizu is now an associate professor at SUNY-Cortland (State University of New York at Cortland) where she is the program manager for the Institute for Geospatial and Drone Technology and the Co-Director of the SUNY Cortland Regional GIS Laboratory.

She found that, following the press announcement on 13 September 2012, traffic on the WMO Extremes website traffic jumped from averaging 150 hits per day before the release to over 24,000 hits over a three-day period after the release. As Dr. Shimizu noted, 'The news story generated intense but relatively short-lived interest with traffic returning to slightly above pre-announcement levels within a few days.' What does that mean? Well, in today's digital media world, it means that scientists making press announcements should prepare for concentrated, but temporary, interest. Sorry, fellow scientists, but don't think that a big science announcement will lead to long-term media interest in your work. For reporters, there is always the next big thing.

Dr. Shimizu also noted that the degree of interest that a science story can generate is likely also tied to the ongoing news cycles. If you issue a science announcement at the same time as some other major news event, it may be that the media coverage of your own event will be lessened. Conversely, another news event may sometimes accent your findings. In the case of our Libyan 58°C press announcement, the gruesome murder of the American ambassador to Libya a few days before our press release likely played a role in the increased public interest of our own announcement about the Libyan temperature.

The geographic extent of the coverage of our Libyan announcement was also interesting. Citizens of Italy comprised the most hits to the WMO extremes site (with the United States, Netherlands and Brazil being the next three). The high number of Italian hits were the result of two things, first that country's historic linkages to Libya and second to a leading Italian news television story highlighting the story in their daily coverage.

Dr. Shimizu and our team also found that a key feature to the global interest in the story was likely the release of the WMO press announcement in six different languages (English, French, Spanish, Chinese, Russian and Arabic). That multi-lingual release enhanced the opportunities for people across the world to get specific details on the WMO evaluation without the filter of secondary translation. She noted that:

> . . . while such linguistic diversity may not be possible in all media releases by atmospheric scientists, researchers may wish to undertake the expense and time to tailor a release in at least a few different languages if the specific nature of the announcement warrants and the opportunity for translations exists.

We found that a significant number of visits to the WMO Extremes Archive webpage originated from other social media sites. Until recently, scientists have not always considered these media, such as Facebook, Google and Twitter (now X), when preparing or making important press announcements of scientific findings. For instance, we found that the Wikipedia site was updated literally seconds after the lightning extremes announcement and that update—including explicit links to the WMO Extremes website—resulted in a number of site visits.

As Dr. Shimizu said:

> At first blush, some might be a bit surprised at the robust response in popular social media to scientific findings. Atmospheric scientists should be aware of the increasingly important disseminating power of the social media.

Finally, she and our team discovered that while many visitors stayed on the site for less than a minute, some visitors remained on the site for extended visits, some as long as ten minutes. She recommended that 'researchers should give thought to not only the specific press release being publicized but to overall website content.' She recommended that careful scrutiny of a science website means that any misspellings, typographical errors, bad link or other errors are identified and corrected *before* making an important announcement. Many academic (and governmental) institutions now have dedicated IT professionals who can assist researchers in that critical web design and maintenance.

While our announcements might serve as an aid to other scientists about how a press announcement is issued, there are also some other groups who have come under our focus. I had mentioned at the end of the last chapter on our Antarctic evaluations that one member of my committee noted that our announcement could also be a nice teachable moment for the media.

In that detailed investigation, we determined that the Seymour Island permafrost monitoring station observation of 20.8°C was invalid because of concerns about the modified radiation shield (see chapter 14 for details). But when news of the observation first became known, global media coverage of the observation as a valid new extreme spread like wildfire.

Our careful analysis shows that media should be cautious in announcing temperature extremes *before* we have thoroughly evaluated those extremes. Unfortunately, we can't produce instantaneous answers of validity—which is what the media and public often demand. In this case, for instance, to achieve the level of absolute accuracy needed for temperature measurements required us to spend a great deal of time studying the screening and radiation shielding. Such detailed analyses are often not appreciated by the media and the public.

To their credit, at the time that the Antarctic observation occurred, many news organizations urged caution. For example, the *Washington Post* reported the potential record observation but noted that the WMO 'is looking into the new report, too, but urged caution about the higher reading' (Freedman 2020).

Many other media outlets did not. For example, the *Guardian* news site initially reported, '[t]he Antarctic has registered a temperature of more than 20° C (68°F) for the first time on record, prompting fears of climate instability in the world's greatest repository of ice' (Watts 2020). I should note that the reporter did later update that initial report to read 'these records will need to be confirmed by the World Meteorological Organization. . . . '

Additionally, news reports first cited a myriad of values for the Antarctic observations. Values of 18.3°C (the accepted temperature) and 18.4°C were reported for the Argentine Esperanza station while most media reported a value of 20.75°C for the Seymour Island permafrost monitoring station (contrary to reporting temperature to the nearest tenth of a degree Celsius). All of this misinformation, in turn, resulted in many social media sites reporting the problematic permafrost station's observation as a truth.

In an attempt to reduce this type of potential misinformation, we at the WMO Archive of Weather & Climate Extremes have instituted a fast response team. With this approach, I assemble within hours or days of a new claim a small (two- or three-person) team of international atmospheric scientists familiar with the type of extreme recorded. Using the best *available* data, that team makes a rapid and very preliminary recommendation whether the extreme is valid.

Following that initial recommendation, the WMO then issues a global press release about the extreme but containing the proviso pending full investigation. A full WMO extreme evaluation team is then created and assembled. That full evaluation team conducts a comprehensive evaluation of the given extreme (including photographs, raw data and metadata of the observation equipment), and we proceed as normal.

One aspect of our press announcements that has been interesting is the greater dependence on visuals. As part of our WMO news releases now, we often have images and maps. Stories are better received when there is accompanying visual information about the extreme. This format often takes the form of actual images (for example, the satellite images of lightning, photographs of the equipment or other aspects of the investigation). Normally, we include a locator map that identifies the exact geographic location of the observation. Since I am a long-time member of a university school of geographical sciences, I have had access to superb cartographers, especially my go-to mapping expert Barbara Trapido-Lurie and her students.

But that sometimes causes minor problems.

I mentioned in the last chapter that, because the WMO is affiliated with the United Nations, we must employ occasionally odd continental definitions, which, for the WMO, are officially labeled, not as continents, but as WMO Administrative Regions. That is also true of international borders. It all boils down to the fact that our cartographic products need to be reviewed to make sure that they are compliant with official UN borders. One can't depend on Google maps in official international work.

Finally, with regard to visuals, I must commend the work of a great class of GIScience students under the direction of Dr. Elizabeth Wentz. When we first constructed the online WMO Extremes website, we didn't have a mapping component. Many of the first messages to me were from people suggesting/demanding that we add an interactive mapping display of the global weather and climate extremes.

Thanks to the work of Dr. Wentz, her talented students, and some gifted consultants, we did exactly that.

One problem that we had to solve in displaying the data was how to create world-recognizable symbols for the various weather extremes. Dr. Wentz's students drafted for our website a set of such international symbols. I was pleased to notice that they had even designed separate tropical cyclone symbols for the Northern and Southern Hemispheres (since such storms rotate in opposite directions in the two hemispheres). The one concession we did make, with regard to the use of English, was in the display of pressure using the letters *H* and *L* to show record high and low pressures. Overall, I believe—and emails have confirmed—that the symbols are indicative of world weather records for a global audience.

*Figure 15.3* Map symbols in WMO Archive of Weather and Climate Extremes Archive.

I must admit that being the lead curator for a global project on weather and climate extremes has given me an education far beyond the limits of evaluating extremes. Translating our work for the media has been one of the most eye-opening challenges. Luckily, I've been able to glean the fundamentals from some of the best experts in the business.

# Interlude: Freaks of the Weather and Media

I mentioned in the preceding chapter that other news events can sometimes detract and limit coverage of a major weather story. As a classic example of that occurred in 1945 when a horrific F5 tornado struck Antler, Oklahoma. The tornado leveled the small town with over six hundred buildings completely destroyed and another seven hundred more damaged. Sixty-nine people were killed including, ironically, the owners of the town's only funeral parlor. An awful disaster! But afterwards, surprisingly little news appeared about the tornado in the newspapers of the time. Without question, this tornado would have received more attention across the country, except that, on that exact day, President Roosevelt died. Indeed, even the newspapers nearby the devastated town issued more information on the death of the President than on the deadly tornado.

As an example of some of the problems associated with rapid reporting, a very destructive hurricane named Bonnie made landfall in North Carolina in 1998. False rumors managed to get into the news about this destructive hurricane. For instance, radio stations reported that both Holden Beach and Caswell Beach were cut off from contact due to surging waters that created new inlets on the barrier islands—neither of which was true. Another rumor that reached all the way to CNN's *Headline News* was that forty-six residents of Bald Head Island had taken refuge inside the island lighthouse. Further reporting revealed that only six residents, four dogs, two cats and a parrot were in the lighthouse—and they all (at least the humans) were there by choice and in no immediate danger.

Yet, even with an occasional misstep, in my humble opinion, broadcast meteorologists hold one of the hardest jobs in the atmospheric sciences. They are the essential conduits who must translate what is, in essence, technical science into understandable public information. Their expert interpretation of the weather situation when relayed to viewers can be in some cases the difference between life and death! Because of their difficult job, those broadcast meteorologists are sometimes held to bizarre standards by their listening audience.

Case in point: In 1996, an Israeli woman sued Haifa's Channel 2 television and its popular weatherman Danny Rup in small claims court for $1,000 after he predicted clear skies and sunshine for a day that turned out to be stormy.

She stated that she dressed too lightly after hearing his assurances of rising temperatures. As a result, she caught the flu, missed four days' work, spent $38 on medication and suffered stress.

Of greater concern, in 2022, two top officials with Hungary's National Meteorological Service (NMS) were summarily fired by the government. They had forecasted that severe storms would impact Budapest on the country's most important national holiday. Instead, those storms passed to the south of the capital.

The storms *missed* Budapest—that's good, right? So what was the problem? The national forecast called for intense storms to occur in the capital around 9 PM local time. That caused the city organizers to postpone a massive annual fireworks display. The fireworks show was to celebrate St. Stephen's Day, a holiday that marks the country's founding, and it is an event that is watched by more than a million people.

After the unsatisfactory forecast, Hungarian media savagely criticized the weather agency. The NMS issued an apology on social media the next day, but it was too late to save the jobs of the agency's chief, Kornelia Radics, and her deputy, Gyula Horvath. Soon after, a number of agency leaders in the Hungarian government posted a statement on social media demanding that their fired colleagues be reinstated as soon as possible. They wrote that the firings were politically motivated and that the forecast was issued based on the best possible information at the time. They stated, 'it is our firm view that, despite considerable pressure from decision-makers, our colleagues . . . provided the best of their knowledge and are not responsible for any alleged or actual damage' (Rosenthal 2023).

'Politically motivated' is a phrase that we hear a lot in today's hyperpolitical world. It is interesting sometimes to delve into history and realize that such political motivation—even about weather—is not a new thing.

In 1853, a respected Prussian meteorologist named Heinrich Wilhelm Dove reported a bizarre linkage of the media, politics . . . and weather . . . in one of the leading scientific journals of the time. Professor Dove recounted the strange episode in 1767 when King Frederick the Great of Prussia ordered that a patently false news account of a massive hailstorm be written solely to deflect people's growing concerns about an impending war.

As Professor Dove wrote:

> A distinguished foreigner visited Berlin in the year 1767, and was invited by Frederick the Great to San Souci.
>
> 'Of what do they talk in Berlin?' asked the King.
>
> 'That your Majesty is arming, and that there will be war,' was the reply.
>
> In order, therefore, to give a different turn to the conversation of the metropolis, the King commanded a report to be drawn up of a severe hailstorm at Potsdam, which was to be copied into the Berlin newspapers, with directions to take no refutation.

> The reporter laid on his colours pretty thick. Masses of ice of the size of a pumpkin had fallen; all the windows had been shattered; a brewer had had his arm broken; and one of the oxen yoked to a wagon had been killed.
>
> On the arrival of the Berlin newspapers at Potsdam—where there had been most beautiful weather during the whole time—astonishment and vexation laid hold of everybody's mind; the neighbourhood rose as one man, seized pen in hand, and protested solemnly to the contrary of what had been stated.
>
> Never had the posts in Berlin received so many letters; each of them asserting that everything was going on as usual in Potsdam, that nothing extraordinary had taken place, no windows had been shattered, no one's arm broken, no living being killed.
>
> But none of these letters were published; the news was copied into all the papers, and the King's design had perfect success. Everywhere the hail-storm, and nothing but the hail-storm, formed the subject of conversation [thereby diverting the public's attention from the impending war]. As the report was never contradicted, it was transcribed into all the scientific compendiums of the day; for at that time people were still possessed by the extraordinary idea that everything contained in a newspaper must be true. (Dove 1853, 222)

Professor Dove demonstrated that the idea of politically-driven climate/weather misinformation isn't—unfortunately—a new one.

# 16 The Future!

> We have to adapt to climate change. That means droughts, flooding, tropical storms, heatwaves, water shortages, coastal inundation.
>
> WMO Secretary-General Professor Petteri Taalas

> "'Data! Data! Data!' [Holmes] cried impatiently. 'I can't make bricks without clay.'"
>
> Arthur Conan Doyle

We live in an increasingly warming world. In recent decades, scientists have attributed that warming, in overwhelming degree, to us—to human activity. As I've said earlier, those are facts, not opinion. For instance, the last coldest record we have accepted into the WMO Archive was for a Greenland temperature occurring in 1992. The latest warmest record extreme we have accepted into the WMO Archive was in 2018, and we are currently evaluating even more recent ones. Yes, we live in a rapidly warming world, an environment that we ourselves are manipulating.

What should we do about it?

A good question, but a question that I personally leave to the experts in such matters. I am *not* a climate policy expert. My professional role is to evaluate the world's weather.

I guess you might describe me as an old-school data analyst. I was taught that scientists should not exceed their personal areas of expertise—or allow their personal opinions to cause others to question the validity of their data. I am not a policy expert. When asked about policy, I worry, because I do not directly work with climate policy, that if I recommend specific policies it could conceivably compromise my scientific integrity. In our politicized and polarized world, the inherent neutrality of the scientific community to politics becomes paramount to the continuing health and survival of our society.

We must be able to trust the science. My job is to ensure that trust. My teams and I analyze the data and let the facts speak for themselves.

'On the job,' I don't like to bicker about politics. I frankly don't care whether my research supports or contradicts any political agenda, party or statement.

But I *love* reasoned scientific debate. That is because thoughtful discussion is a fundamental part of the scientific process.

DOI: 10.4324/9781003367956-16

In this book, I haven't discussed in detail the times when disagreements have broken out in my committees. Such disagreements—scientific disagreements—are critical aspects of science-at-work. My committees pour over the data and evidence and then they *debate* it—sometimes vigorously—until they reach a consensus.

Scientists are human—and so they often have strong ideas about any given subject, particularly one in which they are an expert. In our WMO evaluations, we work through any disagreements until the committee reaches a consensus. Indeed, it is often through disagreements that my committee members uncover critical details that help in our final determination.

But what about other colleagues in the greater scientific community? Do fellow scientists also get a chance to comment on our work?

Yes, even though my committees consist of some of the best climate scientists in the world, I have made a specific point of obtaining broad scientific consensus for our WMO extremes evaluations. Many people might think that having a group of recognized climate experts from around the world check a weather record provides enough justification—enough ammunition, if you will—to stop most arm-chair critics from criticizing our results.

With many organizations like the United Nations, often the formal conclusions of such blue-ribbon committees would be enough.

With the WMO extremes decisions, we have often opted to take our confirmation process one step further. We publish our committee results within science's top academic journals.

In science, for most important research to accepted, it must either be presented at a formal professional conference or it must be published in a professional scientific journal.

Most science disciplines publish monthly technical journals that contain the most important cutting-edge research associated with their profession. In the atmospheric sciences, for example, professional journals include the *Bulletin of the American Meteorological Society*, the *Quarterly Journal of the Royal Meteorological Society*, and the *Australian Meteorological and Oceanographic Journal*, to name a few. Academic journals with broader interests (encompassing several disciplines) include *EOS Earth & Space Science News, Geophysical Research Letters* and the *International Journal of Climatology*. The broadest-based science journals are general multidisciplinary publications such as *Nature* or *Science*.

A key feature of publishing research in a professional journal is that, before publication, articles are subject to a peer-review process. That is when the research article—from which the names of the authors is first extracted—is reviewed by one or more anonymous, external scientists. Such double anonymity helps to ensure that any inherent biases (by authors or reviewers) are minimized.

The journal review process provides a critical second-tier of insurance that the research was evaluated for any flaws. By the way, that holds true with all science stories picked up by the news media. If you see a phrase in the news story reading 'this research has been published in such-and-such scientific journal,' you can give a bit more credence to the research findings than if you don't see such a statement.

The writings on many so-called climate websites or climate blogs, for example, aren't peer-reviewed. There is no external quality control. Especially with regard to climate information, not everything published on the internet is true.

For our WMO evaluations, such extra peer-review validation does take time but, in some cases, I have found that a professional journal's external reviewers have identified interesting features that the committee hadn't fully addressed. That detection thereby improved the research. Afterwards in a few cases when a particularly good reviewer has identified themselves once the review process is completed, I have even later invited them onto future evaluation committees. Fundamentally, formal publication in the top scientific journals adds an extra layer of assurance that our findings are valid.

However, another important aspect of academic publishing is that it also indicates the research's importance to the discipline. Because space in technical journals is limited, the majority of submitted articles are *not* published. Formal publication by those journals also is a measure of the work's relevance to the entire scientific community.

A similar situation exists with presentations at science conferences.

This point was made by one of my frequent committee members (and whom I'm honored to call a friend) Dr. Philip Jones from the Climatic Research Unit at University of East Anglia in Great Britain. Dr. Jones is a very well-known and honored climatologist who (among many other things) is one of the world's foremost authorities on instrumental climate change and the detection of climate change. As I was mulling over ideas for this last chapter (and lamenting about critics), Dr. Jones suggested that I say something about the quality of disagreements at scientific meetings. He has a good point.

In most formal scientific meetings around the globe, there is a question-and-answer period after a presentation. Sometimes these exchanges can become confrontational—as I've noted, scientists can be quite passionate. But—at least in Phil's or my own experiences—the disagreements have been resolved

*Figure 16.1* Left: Dr. Philip Jones, world-renown climatologist (and fellow weather extremes enthusiast). Right: Dr. Kerry Emanuel, world-renown tropical cyclone expert.

Source: Photos courtesy of the University of East Anglia and Dr. Jones and the Department of Earth, Atmospheric and Planetary Sciences at MIT and Dr. Emanuel.

without becoming grievous. Indeed, I've occasionally seen the arguing scientists later in the conference's lounges amicably continuing the discussion over a drink.

It should be noted that disagreements aren't limited to climate science, either. I think that you hear more about climate because we climatologists have become high profile.

Anthropology, archaeology and geology—those are just a few fields in which I myself have seen scientists argue, sometimes strenuously, their science. Exactly when did the first peoples arrive in the Americas? How thoroughly has a sample—like a piece of ancient pottery—been dated? Does a landscape like the Grand Canyon form through periodic flash floods or through stream-capture?

Don't think that those questions are that important? Scientists in those fields would passionately disagree. Yes, scientists indulge in vigorous debate about topics that the general public might consider to be mundane! One of my WMO committee members Dr. Kerry Emanuel phrased it best when he told me, 'Disagreements within science are the norm, and resolving them is part of what drives progress in science.' Dr. Emanuel, a professor at the Massachusetts Institute of Technology, is one of the world's foremost experts on tropical cyclones.

But, as Dr. Phil Jones states, such scientific debate shouldn't lead to war. You shouldn't hate someone because they offer a different interpretation of the facts.

Unfortunately, in the wider world, such is not always the case, particularly with regard to climate and weather. Some climate scientists—many of whom are very good friends of mine—have even opted out of remaining in climate science because of the reaction—even pure hatred—that their work, or the public perception of their work, has produced.

For example, nowadays Dr. Phil Jones and I—and many other scientists in the field of climatology—occasionally receive horrid emails and messages from the public. On social media, I have been called quite offensive names. Our WMO results have been dismissed out of hand as blindly following the political agenda of the United Nations. Over the years, I've learned to live with such venom. But my experiences pale in comparison to some correspondence directed to Dr. Jones throughout his long career of which, he has told me, a small number have stated that he should take his own life.

That is abhorrent!

Yet, still today, I receive messages—even from other academics—saying that I should stop our extremes work because some of your scientists are from 'bad' countries! Or they have told me (often using profane language) that 'some of your scientists have the wrong beliefs [or political viewpoints or affiliations]! You shouldn't even mention them!' I do not enter into debate with these people. Yes, in these modern times, I do realize that my particular field of study, weather and climate science, has become more political and partisan.

Such poor behavior is, as I tell my students, something with which modern climatologists have to deal. 'If you want to remain in academics,' I advise them, 'you must develop a thick skin.'

Yet *reasoned* disagreement and debate are important in the science world.

My WMO committees are composed of diverse, intelligent people from all parts of the world. It should be expected that disagreements will arise. But I am proud to say that in my committees the disagreements have been resolved without personal attacks. Indeed, I have found that good friendships between disagreeing scientists have sometimes resulted after the evaluations are complete!

Whenever the WMO announces a new extreme, two questions that I am asked by journalists is 'Is this is a result of climate change and what should we do about it?'

Yes, those questions are important inquiries. They are deserving of answers. And they are also questions that I defer to the WMO's leader, the Secretary-General, who interprets atmospheric data into definable political actions. That's one of his jobs. At the time of this writing, the WMO's Secretary-General is Professor Petteri Taalas. He is the first person whom I quoted at the beginning of this chapter. Professor Taalas has said many times that 'we have to adapt to climate change. That means droughts, flooding, tropical storms, heatwaves, water shortages, coastal inundation' (Taalas 2022).

That is an admirable position and one that I, as a global citizen, support.

As I have stated, my own personal philosophy is defined by the second quote at the start of this chapter as said by the great fictional detective Sherlock Holmes. '"Data! Data! Data!" [Holmes] cried impatiently. "I can't make bricks without clay."' As the WMO Rapporteur of Weather and Climate Extremes, my primary job is to 'get the facts.' I'll leave the political ramifications of my scientific investigations to other—more talented—scientists such as WMO Secretary-General Taalas.

Okay, with that said as a rationale about why I do not discuss politics, I *do* want to address the future of climate in this final chapter.

When one hears discussions on climate change, the emphasis tends to be focused on the next few decades—and justifiably so. After all, that is the specific time in which we and our children will be living.

But climate change is driven by a variety of different processes created by an assortment of diverse forcing mechanisms. Each of those forcing mechanisms operates at *specific and different* time scales. For example, a single major volcanic eruption, depending on the height and type of eruption, can lead to a pronounced global cooling for one to four years after the eruption ('volcanic climate theory'). The eruption of Mt. Pinatubo caused such a cooling for the first three years of the 1990s.

At a completely different time scale, orbital changes of the earth can also produce changes in global climate. But the time scales of those orbital changes occur over thousands of years ('Milankovitch astronomical climate theory'). I repeat, they affect climate over *thousands* of years. I have seen many social media expert posts claiming that, because the earth has come out of an ice age ten thousand years ago, the current warming of the last century is natural and associated with orbital change. Obviously, such posts don't address how the *time scale* affects climate forcing mechanisms. Milankovitch climate theory—

and, yes, I myself have published scientific work on that theory—does not address change over the time scale of mere *decades*—a time scale covering our lifetimes.

Okay, recent climate change over the last hundred years *cannot* be explained by volcanic climate theory (the changes are too long of a period for the forcing mechanism to influence) or by Milankovitch astronomical climate theory (the changes are far too short for the forcing mechanism to influence), and yet those are only two forcing mechanisms. There are *hundreds* of forcing mechanisms, including such new ideas as the influence of surface and deep-sea ocean flow (for example, professor Wally Broecker's 'global thermohaline circulation climate theory'). Every forcing mechanism operates at a *different* time scale.

So what climate forcing mechanisms *do* influence climate on the time scale of decades—not years or millennia?

The science is clear. Anthropogenic (human-caused) increases in carbon dioxide and other greenhouse gases have repeatably been shown by scientific analyses to be the major influencers of climate over the time scale of decades. Again, I am sorry if that fact offends you.

I understand why such anger occurs. Most of us don't like to be told that our aggregated actions are responsible for a widespread change—damage—that is beyond our direct personal control. I understand that.

Nevertheless, that clear human influence of climate raises one crucial question: Is there anything we can do about it?!

As I mentioned above, I will defer any detailed comments to the actual policy experts such as WMO Secretary-General Taalas who addresses climate strategies. But I will mention one encouraging sign for answering that question—we are becoming more knowledgeable about the climate.

One of my academic titles is 'Distinguished Global Futures Scientist' within the newly created College of Global Futures at Arizona State University. My academic institution was one of the first in the world to create such a program that centers on the earth's environmental future. Now many other universities have such programs.

Those programs all revolve around the concept of sustainability. What is sustainability?

Sustainability has been defined as the field of study that examines the maintenance of human well-being for present and future generations, while safeguarding the planet's life-supporting ecosystems.

My particular university program, geographical sciences, for instance, concentrates on urban climate and how climate impacts a given city's inhabitants. Some of my university colleagues, such as Dr. Ariane Middel, are examining climate details at the microscale. They have developed ways to determine microvariations in city climate. For example, Dr. Middel has engineered small-instrumented rover vehicles that travel over the urban landscape, recording precise temperatures and other weather measurements.

Other scientists at my institution are examining how our human activities might be better adapted to changing climate. For example, Dr. Matei Georgescu has examined, through a set of impressive climate-economic computer simulations, how sustainable biofuels for aviation can be produced in some of the US's marginal agricultural areas. Still another, Dr. Dave Sailor, has studied the effects of recoloring of city roads from black to white to improve local temperature conditions.

These types of study, studying how our environment can be modified, link to a growing area of applied academic work in engineering programs around the world. That area is geoengineering.

Geoengineering—the direct application of engineering to achieve a desired change in climate—is becoming a realistic possibility. More and more, geoengineering is reaching mainstream acceptance. For example, in 2008, Bill Gates, one of the most influential men on the planet, filed patents for a number of geoengineering methods to control and prevent hurricanes.

One method that Gates and his engineers proposed was to deploy fleets of vessels into the Gulf of Mexico. Those ships would pump cold water to the ocean's surface using colder water from great depths. Their reasoning was that, since hurricanes are powered by warm moist air produced by warm waters, cutting off the storm's access to warm water would cause the hurricane to die—before it could hit land and cause death and destruction.

Interesting. Such projects make it sound as if it might one day be possible to *control* weather. To stop weather and climate extremes!

Before we start celebrating, be aware that there is often a fundamental problem with geoengineering. That problem can be encapsulated in one short phrase: unintended consequences.

For example, with the Gates' patent above, marine biologists responded to his idea with the question: what would be the impact of such cold water pumping on the diverse aquatic life found in the Gulf of Mexico? And would those impacts outweigh the advantages of dissipating hurricanes? Those questions point out the difficulty with this technique. Geoengineering methods designed to create one specific change in the climate-environment system may result in a multitude of other changes—unforeseen changes—in that very system.

We have established that our climate system is one of the most complex, interacting systems that we have ever studied. There are no simple answers or quick fixes. The study of climatology can be intimidating.

Most climate scientists acknowledge that the mathematics of our field is hard. An understanding of the nuances—the feedbacks—present in the climate system can be daunting. My college students have to complete four semesters of calculus and two semesters of calculus-based physics *before* they can take certain upper-level university courses in meteorology and climatology. I tell those students that atmospheric science is a unique combination of many disciplines. Those subjects are the most difficult college subjects to master: engineering, physics, mathematics, biology, oceanography, chemistry and many others. Atmospheric science is a challenging subject for many students.

Thus far, I have found that supply has met the demand. Our university atmospheric science program—and many others across the county—continue to attract good, qualified students who are willing to spend the effort and get the necessary background to enter this important field. I've been particularly pleased with my own atmospheric science program which—despite the common perception that the atmospheric sciences are male-oriented—has maintained for years a constant fifty-fifty split of male and female students.

If these students are good indicators, the future of atmospheric sciences is bright.

But what specifically will happen to the WMO Archive of world weather and climate in the future?

First, I can state that the WMO Archive will be expanding, and I base that on the fact that since our creation, we *have been* expanding.

For instance, within the last couple of years, it was brought to our attention that while we had specific world records for the Southern Hemisphere's Antarctic region, there were no equivalent records for the Northern Hemisphere's Arctic region—say, for example, the records at or north of the Arctic Circle, 66.5°N. So, we recently expanded our archive to include the Arctic Circle.

Following that decision, we had to evaluate a potential new record in the summer of 2020 when a maximum temperature of 38°C (100.4°F) was recorded for Verkhoyansk, a town in northeast Russia 115 kilometers north of the Arctic Circle. The WMO and I created an ad-hoc international panel of atmospheric scientists tasked with analysis and verification of Arctic extreme temperatures.

That investigation also points out another continuing aspect of this work: politics will continue to be involved. As we commenced that Verkhoyansk Russia investigation, we—and our Russian committee members—had to contend with the unfortunate political concerns associated with Russia's incursion into Ukraine. As I mentioned, politics has intruded more and more into science.

But the news isn't all bad. The future of the WMO Archive of Weather and Climate Extremes does hold some interesting possibilities.

For example, I've had some scientists approach me suggesting that we should consider expanding our WMO Archive to other planets. Since exploration of Mars has begun, space probes such as Viking, Mars Pathfinder, Phoenix, the Curiosity rover and others have made direct weather measurements. The probes take weather observations just like their terrestrial counterparts.

Most Martian weather stations have used thermocouples or thermocapacitors for air temperature and hot-film sensors to measure winds. For the Martian thin atmosphere (only a thousandth of our own) pressure sensors are based on flexing diaphragms, although more recently miniature sensors have been used which are similar to those developed for earth-based radiosondes (discussed in chapter 11). Humidity is also now measured by miniature sensors of the type developed for terrestrial radiosondes.

Given that increasing amount of extraterrestrial weather information, should we tally—and verify—Martian weather extremes?

It is a question that we are discussing.

Whatever happens to the archive, I have learned to be flexible. As climate changes, so will the WMO Archive of climate and weather extremes.

To complete the circle that I started in the first chapter, the necessity of knowing how much climate is changing was one of the primary reasons why I created the WMO Extremes Archive. We need to know—and be confident about—our *existing* weather and climate extremes to determine how much and how fast our world's climates are changing. In other words, a good understanding of our current extremes establishes the critical baselines that we need to access how our climate is changing.

Over the past nearly twenty years, we have strived at the WMO Extremes Archive to achieve that.

Yes, the future will be different than the present. But no matter how different the climatic future may be for humanity compared to its past there is one major counteracting factor that I have discovered over the last twenty years of this WMO project that may tip the balance in our favor.

What is that factor?

You may have noticed that, throughout all investigations that the World Meteorological Organization has undertaken since the start of the Archive of Climate and Weather Extremes, there has been one central, unifying aspect that led to the success of those investigations. That feature is this: each team was composed of a set of brilliant scientists—experts of all ages, genders, ethnicities, faiths and commitments—who accepted responsibility, worked together and then produced extraordinary results.

I have tried to highlight some of the stories of a few of these incredible people and their inspiring work through this book.

I belief that our future mitigation of climate change—and the ultimate direction of humanity's future—rests on the shoulders of these, and many other, brilliant scientists and engineers as well as the upcoming future generations of scientists. I am optimistic that as long as these learned people have the ability to do their work—and as long as the public *respects* these people's genius enough to carry out that work—that we can find solutions to the problems posed by ongoing decadal climate change.

Yes, it won't be easy . . . but it will be possible.

The efforts of myself and World Meteorological Organization scientists presented here have been small steps working towards that solution. What Vincent van Gogh said in 1882 is still true today:

> For the great doesn't happen through impulse alone, and is a succession of little things that are brought together.

# References

## Chapter 1 and Interlude

Cerveny, Randy. 2005. *Freaks of the Storm, from Flying Cows to Stealing Thunder: The World's Strangest True Weather Stories*. New York:Basic Books.

Cerveny, R. S., J. Lawrimore, R. Edwards, C. Landsea. 2007. 'Extreme Weather Records: Compilation, Adjudication and Publication.' *Bulletin of the American Meteorological Society*, 88, no. 6: 853–860.

DeBiasse, Kimberly. 2001. *Acclimation's Influence on Physically-Fit Individuals: Marathon Race Results as a Function of Meteorological Variables and Indices*. Ph.D. diss. Arizona State University.

Henry, O. 1997. 'A Fog in Santone' In *O. Henry: 100 Selected Short Stories*. Ware: Wordsworth Editions Ltd.

Lane, Frank W. 1965. *The Elements Rage*. Philadelphia, PA:Chilton Book Co.

Larsen, Erik. 2011. *Isaac's Storm: A Man, a Time, and the Deadliest Hurricane in History*. New York: Vintage Press.

Lawai, Nurudeen. 2021. '"Mysterious" Cloth Fell from the Sky in Ondo? Truth Finally Emerges.' *Legit*. Accessed 11 October 2023. https://www.legit.ng/nigeria/1442962-mysterious-cloth-fell-sky-ondo-truth-finally-emerges/.

National Oceanic and Atmospheric Administration (NOAA). 2023. National Climate Extremes Committee. Accessed 10 September 2023. https://www.ncei.noaa.gov/access/monitoring/ncec/.

Phillips, David. 2000. *Blame it on the Weather: Strange Canadian Weather Facts*. Toronto: Key Porter Books Ltd.

*UK Independent*. 2020. 'Chocolate snow falls on Swiss town after ventilation defect at Lindt factory'https://www.independent.co.uk/news/world/europe/chocolate-snow-lindt-factory-switzerland-ventilation-system-a9675611.html.

US Army Corps of Engineers Topographic Engineering Center. 1996. *Weather and Climate Extreme*, by Paul F.Krause and Kathleen L.Flood. TEC-0099. Alexandria, VA: US Army Corps of Engineers.

'World Meteorological Organization Home Page.' World Meteorological Organization. Accessed 6 October 2023. https://public.wmo.int/en.

'World Weather and Climate Extremes Archive.' World Meteorological Organization. Accessed 27 September 2023.https://wmo.asu.edu/.

**Chapter 2 and Interlude**

Anonymous. 1856. 'Curiosities of the Thunder-Storms.' *Eclectic Magazine* 38: 458–471.

Cerveny, R. S. 1994. 'Power of the Gods.' *Weatherwise* 47, no. 2: 1023.

Cerveny, Randy. 2019. 'Strange Bedfellows: The Long History of Politics and Meteorology.' *Weatherwise* 72, no.1(2018): 30–35. doi:10.1080/00431672.2019.1538763.

Graham, Steve. 'Vilhelm Bjerknes.'NASA: Earth Observatory. Accessed October 6, 2023. https://earthobservatory.nasa.gov/features/Bjerknes.

'History of WMO.'World Meteorological Organization. 18 January 2023. https://public.wmo.int/en/about-us/who-we-are/history-of-wmo.

Jerome, Jerome K. *Idle Ideas in 1905*. London:Hurst & Blackett, 1905. Available at: https://www.gutenberg.org/ebooks/3140.

Loebsack, Theo. 1959. *Our Atmosphere*. Translated by E. L. and D. Rewald. New York: Pantheon.

Merriam-Webster. 2023. 'Rapporteur.' Accessed 10 September 2023. https://www.merriam-webster.com/dictionary/rapporteur.

'World Meteorological Organization Home Page.'World Meteorological Organization. Accessed 6 October 2023. https://public.wmo.int/en.

'World Weather and Climate Extremes Archive.'World Meteorological Organization. Accessed 27 September 2023.https://wmo.asu.edu/.

**Chapter 3 and Interlude**

Cerveny, R., V. D.Belitskaya, P.Bessemoulin, M.Cortez, C.Landsea and T. C. Peterson, 'A New Western Hemisphere 24-Hour Rainfall Record.' *WMO Bulletin* 56, no. 3 (2007): 212–215.

Cerveny, R. S. and M. G. Marcus. 1994. 'Elements of Espionage.' *Weatherwise* 47, no. 5 (2010): 14–21.

'Faces of the National Weather Service.'National Weather Service, 14 September 2023. https://www.weather.gov/careers/meteorology.

Pasch, Richard J., Eric S. Blake, Hugh D. Cobb III, and David P. Roberts. 'Tropical Cyclone Report: Hurrican Wilma.'National Hurricane Center and and Central Pacific Hurricane Center. Accessed 6 October 2023. https://www.nhc.noaa.gov/ms-word/TCR-AL252005_Wilma.doc.

'World Meteorological Organization Home Page.' World Meteorological Organization. Accessed 6 October 2023. https://public.wmo.int/en.

'World Weather and Climate Extremes Archive.' World Meteorological Organization. Accessed 27 September 2023. https://wmo.asu.edu/.

**Chapter 4 and Interlude**

Blagden, Charles. 1775. 'Experiments and Observations in an Heated Room by Charles Blagden, M.D.F.R.S.' *Philosophical Transactions (1683–1775)* 65: 111–123.

Brunk, I. W. 1958. 'Weather Rambles: Weather Sayings.' *Weatherwise* 11, no. 6: 204–205.

Lee, Douglas H. K. 1948. 'Physiological Climatology.' Class Notes, Douglas H. K. Lee Papers Collection UQFL538. The Fryer Library. University of Queensland, Brisbane, Australia. https://manuscripts.library.uq.edu.au/uqfl538.

El Fadli, K., R. S.Cerveny, C. C.Burt, P.Eden, D.Parker, M.Brunet, T. C.Peterson, et al. 2013. 'World Meteorological Organization Assessment of the Purported World Record 58ERRORC Temperature Extreme at El Azizia, Libya (13 September 1922).'

*Bulletin of the American Meteorological Society* 94, no. 2:199–204. doi:10.1175/BAMS-D-12-00093.1.

Kawczynski, Daniel. 2011. *Seeking Gaddafi: Libya, the West and the Arab Spring*. London: Biteback.

'World Meteorological Organization Home Page.' World Meteorological Organization. Accessed October 6, 2023. https://public.wmo.int/en.

'World Weather and Climate Extremes Archive.' World Meteorological Organization. Accessed September 27, 2023. https://wmo.asu.edu/.

### Chapter 5 and Interlude

Anonymous. 1866. 'General News.' *The Pittsburgh Gazette*: 3,column1.

Cerveny, Niccole, RobertC.Balling Jr., and Randy Cerveny. 2022. 'Dam Weather: The Surprising Impact of Weather on Hoover Dam.' *Weatherwise* 75, no. 5: 42–49. doi:10.1080/00431672.2022.2087441.

Court, Arnold. 1949. 'How Hot Is Death Valley?' *Geographical Review* 39, no. 2: 214–220.

El Fadli, K., R. S. Cerveny, C. C. Burt, P. Eden, D. Parker, M. Brunet, T. C. Peterson, G. Mordacchini, V. Pelino, P. Bessemoulin, J. L. Stella, F. Driouech, M. M Abdel Wahab, M. B. Pace. 2013. 'World Meteorological Organization Assessment of the Purported World Record 58°C Temperature Extreme at El Azizia, Libya (13 September 1922).' *Bulletin of the American Meteorological Society* 94, no. 2: 199–204. doi:10.1175/BAMS-D-12-00093.1.

Merlone, A., H. Al-Dashti, N. Faisal, R. S. Cerveny, S. AlSarmi, P. Bessemoulin, M. Brunet, et al. 2019. 'Temperature Extreme Records: World Meteorological Organization Metrological and Meteorological Evaluation of the 54.0°C Observations in Mitribah, Kuwait and Turbat, Pakistan in 2016/2017.' *International Journal of Climatology* 39, no. 13: 5154–5169. doi:10.1002/joc.6132.

Middleton, W. E Knowles. 1966. *A History of the Thermometer and Its Uses in Meteorology*. Baltimore: The John Hopkins Press.

'World Meteorological Organization Home Page.' World Meteorological Organization. Accessed 6 October 2023. https://public.wmo.int/en.

'World Weather and Climate Extremes Archive.' World Meteorological Organization. Accessed 27 September 2023. https://wmo.asu.edu/.

### Chapter 6 and Interlude

Anonymous. 1930. 'Historic Natural Events.' *Nature* 125: 73.

Anonymous. 1848. 'Windfall.' *Scientific American* 3, no. 15: 113.

Cerveny, R. S. 2005. '*Winds and Wind Systems.' In Encyclopedia of World Climatology, ed.* J. E. Oliver. Dordrecht:Springer Press.

Courtney, Joe, SteveBuchan, Randall S.Cerveny, Pierre Bessemoulin, Thomas C.Peterson, Jose M.Rubiera Torres, John Beven, JohnKing, Blair Trewin, and Kenneth Rancourt. 2012. '*Documentation and Verification of the World Extreme Wind Gust Record: 113.3 m $s^{-1}$ on Barrow Island Australia during Passage of Tropical Cyclone Olivia.' Australian Meteorological and Oceanographic Journal* 62, no.1: 1–9. doi:10.22499/2.6201.001.

Flammarion, C. 1901. *The Unknown (L'inconnu)*. New York:Harper and Brothers.

Lewis, C. S. 2009. *Mere Christianity*. New York:HarperOne.

Putnam, WilliamLowell. 1991. *The Worst Weather on Earth: A History of the Mount Washington Observatory*. Gortham, NH:American Alpine Club.
'Support Mount Washington Observatory.' Mount Washington Observatory. Accessed 2 October 2023. https://www.mountwashington.org/get-involved/support-the-obs/.
'Windfall.'Merriam-Webster. Accessed 6 October 2023.https://www.merriam-webster.com/dictionary/windfall.
"World Meteorological Organization."World Meteorological Organization. Accessed 27 September 2023. https://public.wmo.int/en.
"World Meteorological Organization's World Weather & Climate Extremes Archive."-World Meteorological Organization. Accessed 2 October 2023. https://wmo.asu.edu/.

**Chapter 7 and Interlude**

'Community Collaborative Rain, Hail & Snow Network.' CoCoRaHS. Accessed 6 October 2023. https://www.cocorahs.org/.
Geddes, A. E. M. 1930. *Meteorology. An Introductory Treatise*. London: Blackie & Son Limited.
Guttman, Nathaniel B. 1989. 'Statistical Descriptors of Climate.' *Bulletin of the American Meteorological Society* 70: 602–607.
Hering, D. W. 1924. *Foibles and Fallacies of Science*. New York: Van Nostrand.
Livingston, David. 2006. *Missionary Travels and Researches in South Africa*, Urbana, IL: Project Gutenberg. Accessed 6 October 2023. https://www.gutenberg.org/ebooks/1039.
Strangeways, Ian. 2000. *Measuring the Natural Environment*. Cambridge: Cambridge University Press.
Trewin, Blair. 2017. *WMO Guidelines on the Calculation of Climate Normals*, WMO-No. 1203. Geneva: World Meteorological Organization.
'World Meteorological Organization Home Page.' World Meteorological Organization. Accessed 6 October 2023. https://public.wmo.int/en.
'World Weather and Climate Extremes Archive.' World Meteorological Organization. Accessed 27 September 2023. https://wmo.asu.edu/.
World Meteorological Organization. 2014. 'Cherrapunji, India, Holds New Record for 48-Hour Rainfall.' Accessed 10 October 2023. https://public.wmo.int/en/media/press-release/no-988-cherrapunji-india-holds-new-record-48-hour-rainfall.

**Chapter 8 and Interlude**

Bierce, Ambrose. 1911. *The Devil's Dictionary*. Urbana, Illinois: Project Gutenberg. Accessed 6 October 2023. https://www.gutenberg.org/ebooks/972.
Purevjav, G., R. C. Balling, R. S. Cerveny, R. Allan, G. P. Compo, P. Jones, T. C. Peterson, et al. 2015. '*The Tosontsengel Mongolia World Record Sea-Level Pressure Extreme: Spatial Analysis of Elevation Bias in Adjustment-to-Sea-Level Pressures.*' *International Journal of Climatology* 35, no. 10: 2968–2977. doi:10.1002/joc.4186.
Hess, Seymour. 1959. *Introduction to Theoretical Meteorology*. New York: Henry Holt.
Calandra, Dr.Alexander. 1964. 'The Barometer Story: A Problem in Teaching Critical Thinking.' *Current Science* 49.
'World Meteorological Organization Home Page.' World Meteorological Organization. Accessed 6 October 2023. https://public.wmo.int/en.
'World Weather and Climate Extremes Archive.' World Meteorological Organization. Accessed 27 September 2023. https://wmo.asu.edu/.

**Chapter 9 and Interlude**

Anonymous. 2023. 'Tales from our Archives: Wilson A. Bentley: Pioneering Photographer of Snowflakes.' *Scientific American*. Accessed 10 September 2023. https://siarchives.si.edu/history/featured-topics/stories/wilson-bentley-pioneering-photographer-snowflakes.

Bentley, Wilson A. 1903. *Studies Among the Snow Crystals during the Winter of 1901–2 with Additional Data Collected during Previous Winters and Twenty-Two Half-Tone Plates in Annual Summary of the Monthly Weather Review for 1902*. Washington, DC: US Government Printing Office.

Cerveny, Randy, CharlesKnight, and Nancy Knight. 2005. 'Strange Tales of Hail.' *Weatherwise* 58, no. 3: 28–33.

Dove, HeinrichWilheim. 1853. 'Meteorological Phenomena in connection with the Climate of Berlin.' *Edinburgh Philosophical Journal* 54, no. 108: 222–223.

Kepler, Johannes. 2011. *The Six-Cornered Snowflake*, translated by Jacques Bromberg. Philadelphia: Paul Dry Books.

Thoreau, HenryDavid. 2011. *The Journal of Henry David Thoreau, 1837–1861*. New York: Review of Books.

'Wilson A. Bentley: Pioneering Photographer of Snowflakes.' Smithsonian Institution Archives. 14 September2012. https://siarchives.si.edu/history/featured-topics/stories/wilson-bentley-pioneering-photographer-snowflakes.

'Record Setting Hail Event in Vivian, South Dakota on July 23, 2010.' National Weather Service. Accessed 6 October 2023. https://www.weather.gov/abr/vivianhailstone.

'World Meteorological Organization Home Page.' World Meteorological Organization. Accessed 6 October 2023. https://public.wmo.int/en.

'World Weather and Climate Extremes Archive.' World Meteorological Organization. Accessed 27 September 2023. https://wmo.asu.edu/.

**Chapter 10**

Alexander, W. H. 1902. *Hurricanes: Especially those of Porto Rico and St. Kitts (US Department of Agriculture)*. Weather Bureau, Bulletin No. 32. Washington, DC: US Government Printing Office, 79.

The American Bird Conservatory. 2014. *Hurricanes Present Another Threat to Birds Before and During Fall Migration*. Accessed 6 October 2023. https://abcbirds.org/news/hurricanes-present-another-threat-to-birds-before-and-during-fall-migration/.

American Meteorological Society (AMS). 2018. 'Chris Landsea.' Oral History Project Archives. Accessed 19 September 2023. https://www.ametsoc.org/index.cfm/ams100/oral-histories/chris-landsea/.

Blanc, M. L., and E. Carson. 1948. Finding and Tracking the Hurricane. *Weatherwise* 1: 73–75.

Cerveny, Randy. 2007. 'Weather and Mongols—How the Forces of Nature Helped Shape an Empire.' *Weatherwise* 60: 23–27.

Douglas, Marjory Stoneman. 1958. *Hurricane*. New York: Rinehart & Company.

Enloe, Jesse. 2010. 'Vivian_Hail_Stone_July232010.' NOAA's National Climatic Data Center, Asheville, NC.

Fritz, Angela. 2015. 'World Record?: 100 Inches of Snow May Have Clobbered Italy in 18 Hours, Review Pending.' *Washington Post* (11 March). Accessed 10 September 2023. https://www.washingtonpost.com/news/capital-weather-gang/wp/2015/03/11/100-inches-of-snow-may-have-clobbered-italy-in-18-hours-review-pending/.

Froissart, SirJohn. 1961. *The Chronicles of England, France and Spain*. New York: Dutton & Co.

Garriott, E. B. 1909. 'Weather, Forecasts, and Warnings.' *Monthly Weather Review* (October): 829–830.

Gentry, Robert C. 1964. 'The 1954 Hurricane Season.' *Weatherwise*, 8:1, 12–17.

Libbrecht , Kenneth G. 2007. 'The Formation of Snow Crystals: Subtle Molecular Processes Govern the Growth of a Remarkable Variety of Elaborate Ice Structures.' *American Scientist* 95: 52–59.

Mozai, Tarao. 1982. 'The Lost Fleet of Kublai Khan.' *National Geographic Magazine* 162: 634–648.

National Oceanic and Atmospheric Administration (NOAA). 1997. 'Evaluation of the Reported January 11-12, 1997, Montague, New York, 77-Inch, 24-Hour Lake-Effect Snowfall.' US Department of Commerce. Accessed 10 October 2023. https://www.ncei.noaa.gov/monitoring-content/extremes/ncec/reports/national-seasonal-snowfall-record.pdf.

Spignesi, Stephen J. 1994. *The Odd Index, the Ultimate Compendium of Bizarre and Unusual Facts*. New York: Plume Books.

Tannehill, Ivan R. 1944. *Hurricanes (their Nature and History, particularly those of the West Indies and the Southern Coasts of the United States)*. Princeton, NJ: Princeton University Press.

Woodruff, J. D., K. Kanamaru, S. Kundu and T. L. Cook. 2015. 'Depositional Evidence for the Kamikaze Typhoon and Links to Changes in Typhoon Climatology.' *Geology* 43: 91–94 doi:10.1130/G367209.

World Meteorological Organization (WMO). 2023. 'Measurement of Cryospheric Variables.' *CIMO Guide to Instruments and Methods of Observation—No. 8*, vol. II. Accessed 10 October 2023. https://community.wmo.int/en/activity-areas/imop/wmo-no_8,

'World Meteorological Organization Home Page.' World Meteorological Organization. Accessed 6 October 2023. https://public.wmo.int/en.

'World Weather and Climate Extremes Archive.' World Meteorological Organization. Accessed 27 September 2023. https://wmo.asu.edu/.

**Chapter 11 and Interlude**

*American Experience*. 2020. Season 32, episode 8, 'Mr. Tornado.' Aired May 19. PBS. https://www.pbs.org/video/mr-tornado-upsuu9/.

Brooks, H. E. 2004. 'On the Relationship of Tornado Path Length and Width to Intensity.' *Weather Forecasting* 19:310–319.

Edwards, Roger. 2023. 'The On-Line Tornado FAQ.' Storm Prediction Center. Accessed 6 October 2023. https://www.spc.noaa.gov/faq/tornado/

Ludlum, D. M. 1970. *Early American Tornadoes: 1567–1870*. American Meteorological Society.

Faber, Madeline. 2018. 'Tornado in Virginia Cracks Open Tree Filled with 70,000 Bees: "It Was a Catastrophic Situation,"' September 20. *Fox News*. Accessed 6 October 2023. http://www.foxnews.com/science/2018/09/20/tornado-in-virginia-cracks-open-tree-filled-with-70000-bees-it-was-catastrophic-situation.html.

Grazulis, Thomas P. 1993. *Significant Tornadoes 1680–1991*. St. Johnsbury, VT: Environmental Films Publishing.

National Centers for Environmental Information (NCEI). 2023. 'Storm Event Database: 31 March 2006 F2 Tornado.' Accessed 10 September 2023. https://www.ncdc.noaa.gov/stormevents/eventdetails.jsp?id=5493251.

National Oceanic and Atmospheric Administration (NOAA). 2023. 'The Historic Forecast.' Accessed 10 September 2023. https://www.outlook.noaa.gov/tornadoes/torn50.html.

National Oceanic and Atmospheric Administration Storm Prediction Center (NOAA SPC). 2023. 'Fujita Damage Scale.' Accessed 10 September 2023. https://www.spc.noaa.gov/faq/tornado/f-scale.html Accessed 9/10/2023.

NWS Norman Oklahoma. 2023. 'The May 31–June 1, 2013 Tornado and Flash Flooding Event.' Accessed 10 September 2023. https://www.weather.gov/oun/events-20130531.

Wagler, D. 1966. *The Mighty Whirlwind.* Ontario: Pathway Publishing.

Wagner, Melissa, Robert K.Doe, Aaron Johnson, ZhiangChen, Jnaneshwar Das, and Randall S.Cerveny. 2019. 'Unpiloted Aerial Systems (UASs) Application for Tornado Damage Surveys: Benefits and Procedures.' *Bulletin of the American Meteorological Society* 100: 2405–2409.doi:10.1175/BAMS-D-19-0124.1.

'World Meteorological Organization Home Page.' World Meteorological Organization. Accessed 6 October 2023. https://public.wmo.int/en.

'World Weather and Climate Extremes Archive.' World Meteorological Organization. Accessed 27 September 2023. https://wmo.asu.edu/.

Wurman, Joshua, KarenKosiba, Paul Robinson and Tim Marshall. 2014. 'The Role of Multiple-Vortex Tornado Structure in Causing Storm Researcher Fatalities.' *Bulletin of the American Meteorological Society* 95: 31–45. doi:10.1175/BAMS-D-13-00221.1.

## Chapter 12 and Interlude

Duncombe, J. 2021. 'Have You Seen Ball Lightning? Scientists Want to Know about It.' *Eos* 102 (15 July). doi:10.1029/2021EO160794.

Lang, Timothy J., Stéphane Pédeboy, WilliamRison, Randall S.Cerveny, Joan Montanyà, SergeChauzy, Donald R.MacGorman, et al. 2017. 'WMO World Record Lightning Extremes: Longest Reported Flash Distance and Longest Reported Flash Duration.' *Bulletin of the American Meteorological Society* 43, no. 16: 1153–1168. doi:10.1175/BAMS-D-16-0061.1.

Østgaard, N., T. Gjesteland, B. E. Carlson, A. B. Collier, S. A. Cummer, G. Lu, and H. J. Christian. 2013. 'Simultaneous Observations of Optical Lightning and Terrestrial Gamma Ray Flash from Space.' *Geophysical Research Letters* 40: 2423–2426. doi:10.1002/grl.50466.

Peterson, M. J., T. J. Lang, E. C. Bruning, R. Albrecht, R. J. Blakeslee, W. A. Lyons, S. Pédeboy, et al. 2020. 'New WMO Certified Megaflash Lightning Extremes for Flash Distance (709 km) and Duration (16.73 seconds) Recorded from Space.' *Geophysical Research Letters Geophysical Research Letters* 47, no. 16. doi:10.1029/2020GL088888.

Peterson, Michael J., Timothy J.Lang, Timothy Logan, CheongWee Kiong, MorneGijben, Ron Holle, Ivana Kolmasova, et al. 2022. 'New WMO Certified Megaflash Lightning Extremes for Flash Distance and Duration Recorded from Space.' *Bulletin of the American Meteorological Society* 103, no. 4: 257–261. doi:10.1175/BAMS-D-21-0254.1.

Tomlinson, Charles. 1860. *The Thunder-Storm: An Account of the Properties of Lightning and of Atmospheric Electricity in Various Parts of the World.* London: Society for Promoting Christian Knowledge.

Twain, Mark. 2014. 'Letter to Henry W. Ruoff, 28 August 1908.' In *The Complete Letters of Mark Twain,* volume 6. E-Artnow Publishing.

'World Meteorological Organization Home Page.' World Meteorological Organization. Accessed 6 October 2023. https://public.wmo.int/en.

'World Weather and Climate Extremes Archive.' World Meteorological Organization. Accessed 27 September 2023. https://wmo.asu.edu/.

### Chapter 13 and Interlude

Anonymous. 1888. 'Hail.' *Nature* 37: 42.

'African Centres for Lightning and Electromagnetics Network.' African Centres for Lightning and Electromagnetics Network. Accessed 6 October 2023.https://ACLENet.org.

Cerveny, Randall S., Pierre Bessemoulin, Christopher C.Burt, MaryAnn Cooper, M. D.Zhang Cunjie, AshrafDewan, Jonathan Finch, et al. 2017. 'WMO Assessment of Weather and Climate Mortality Extremes: Lightning, Tropical Cyclones, Tornadoes, and Hail.' *Journal of Weather and Climate in Society* 9, no. 3:487–497. doi:10.1175/WCAS-D-16-0120.1.

Guterres, António. 2020. 'The State of the Planet: Secretary-General's Address at Columbia University.' United Nations. Accessed 6 October 2023. https://www.un.org/sg/en/content/sg/speeches/2020-12-02/address-columbia-university-the-state-of-the-planet.

'Weather Safety Rules.' National Weather Service (June 2). Accessed 6 October 2023. https://www.weather.gov/lmk/weathersafetyrules.

'World Meteorological Organization Home Page.' World Meteorological Organization. Accessed 6 October 2023. https://public.wmo.int/en.

'World Weather and Climate Extremes Archive.' World Meteorological Organization. Accessed 27 September 2023. https://wmo.asu.edu/.

### Chapter 14 and Interlude

Laska, Kamil, J.King, D.Bromwich, P.Jones, S.Solomon, J.Renwick, M.Lazzara, et al. 2018. 'Antarctic Extreme Temperature Record Evaluation.' World Meteorological Organization. Accessed 6 October 2023. https://public.wmo.int/en/resources/meteoworld.

'The Montreal Protocol.' 2023. United NationsEnvironmental Programme. https://www.unep.org/ozonaction/who-we-are/about-montreal-protocol.

'Peaceful Use, Freedom of Scientific Investigation and Cooperation.' Secretariat of the Antarctic Treaty. Accessed 6 October 2023. https://www.ats.aq/index_e.html.

Scott, Robert Falcon. 2008. 'Diary Entry for 17 January 1921.' In *Journals: Captain Scott's Last Expedition*. Oxford: Oxford University Press.

Skansi, Maria de Los Milagros, John King, Matthew A.Lazzara, Randall S.Cerveny, JoseLuis Stella, Susan Solomon, Phil Jones, et al. 2017. 'Evaluating Highest Temperature Extremes for the Antarctic Region.' *EOS* (1 March). doi:10.1029/2017EO068325.

Solomon, Susan. 2001. *The Longest March: Scott's Fatal Antarctic Expedition*. New Haven: Yale University Press.

Stevens, Michael H., Cora E.Randall, Justin N.Carstens, David E.Siskind, John P. McCormack, David D.Kuhl, and Manbharat S.Dhadly. 2022. 'Northern Mid-Latitude Mesospheric Cloud Frequencies Observed by AIM/CIPS: Interannual Variability Driven by Space Traffic.' *Earth and Space Science* 9, no 6. doi:10.1029/2022EA002217.

Thomas, Vanessa and Miles Hatield. 2022. 'Rocket Launches Can Create Night-Shining Clouds Away from the Poles.' NASA (July 21). https://www.nasa.gov/feature/goddard/2022/sun/rocket-launches-can-create-night-shining-clouds-away-from-poles-nasa-aim-mission.

United Nations Environment Programme (UNEP). 2023. 'About Montreal Protocol.' Accessed 10 September 2023. https://www.unep.org/ozonaction/who-we-are/about-montreal-protocol.

'World Meteorological Organization Home Page.' World Meteorological Organization. Accessed 6 October 2023. https://public.wmo.int/en.

'World Weather and Climate Extremes Archive.' World Meteorological Organization. Accessed 27 September 2023. https://wmo.asu.edu/.

### Chapter 15 and Interlude

Barnes, J. 2001. *North Carolina's Hurricane History*. Chapel Hill: The University of North Carolina Press.

Disraeli, Benjamin. 1871. 'Third Reading.' House of Commons Debate208, cc1096-145 (8 August). Accessed 6 October 2023. https://api.parliament.uk/historic-hansard/commons/1871/aug/08/third-reading#S3V0208P0_18710808_HOC_40

Dove, Heinrich Wilhelm. 1853. 'Meteorological Phenomena in Connection with the Climate of Berlin.' Translated by Mrs. Ann Ramsden Bennett from the German of Professor Dove. *The Edinburgh New Philosophical Journal* 54: 214–228.

Freedman, Andrew. 2020. 'Temperature in Antarctica Soars to Near 70 Degrees, Appears to Toppled Continental Record Set Days Earlier.' *The Washington Post*. Accessed 10 September 2023. https://www.washingtonpost.com/weather/2020/02/13/antarctica-hottest-temperature-70-degrees/.

Grazulis, Thomas P. 1993. *Significant Tornadoes*. St. Johnsbury, VT: Environmental Films Publishing.

Rosenthal, Zach. 2022. 'Top Hungarian Weather Service Officials Fired after Wrong Forecast.' *Washington Post* (23 August). Accessed 6 October 2023.https://www.washingtonpost.com/climate-environment/2022/08/23/weather-hungary-meteorologists-fired/.

Shimizu, Melinda, Randall S. Cerveny, Elizabeth A. Wentz, Kevin E. McHugh. 2014. 'Geographic and Virtual Dissemination of an International Climatic Announcement.' *Bulletin of the American Meteorological Society* 95, no. 7: 987–989.

Watts, Jonathan. 2020. 'Antarctic Temperature Rises above 20C for First Time on Record.' *The Guardian*. Accessed 10 September 2023. https://www.theguardian.com/world/2020/feb/13/antarctic-temperature-rises-above-20c-first-time-record.

'World Meteorological Organization Home Page.' World Meteorological Organization. Accessed 6 October 2023. https://public.wmo.int/en.

'World Weather and Climate Extremes Archive.' World Meteorological Organization. Accessed 27 September 2023. https://wmo.asu.edu/.

### Chapter 16

Doyle, Arthur Conan, 1999: The Adventures of Sherlock Holmes, Urbana, IL: Project Gutenberg. Accessed 12 October 2023. https://www.gutenberg.org/ebooks/1661.

Sterin, Alex, PhillipJones, Blair Trewin, Daniel Krahenbuhl and Randall S.Cerveny. 2022. 'Chronicling the Hottest, Coldest, Windiest, and Rainiest Weather.' *EOS Earth & Space Science News* (9 May). Accessed 6 October 2023. doi:10.1029/2022EO220148.

Taalas, WMO Secretary-GeneralPetteri. 2022. 'Opening Remarks at the opening of the IPCC55.' Intergovernmental Panel on Climate Change (IPCC). Accessed 6 October 2023. https://www.ipcc.ch/site/assets/uploads/2022/02/IPCC55-opening-remarks-WMO-SG.pdf.

van Gogh, Vincent, 2023: Letter to brother Theo van Gogh. The Hague, Sunday, 22 October 1882. The Van Gogh Letters organization. https://vangoghletters.org/vg/letters/let274/letter.html (accessed 4/10/2023).

'World Meteorological Organization Home Page.' World Meteorological Organization. Accessed 6 October 2023. https://public.wmo.int/en.

'World Weather and Climate Extremes Archive.' World Meteorological Organization. Accessed 27 September 2023. https://wmo.asu.edu/.

# Index

For Product Safety Concerns and Information please contact our EU representative GPSR@taylorandfrancis.com
Taylor & Francis Verlag GmbH, Kaufingerstraße 24, 80331 München, Germany

www.ingramcontent.com/pod-product-compliance
Lightning Source LLC
LaVergne TN
LVHW011817200926
848484LV00005B/1220

* 9 7 8 1 0 3 2 4 3 5 7 0 1 *